# FIELD-ION MICROSCOPY

*SERIES*    DEFECTS IN CRYSTALLINE SOLIDS

Editors:

S. Amelinckx
R. Gevers
J. Nihoul

Studiecentrum voor kernenergie, Mol,
and
University of Antwerpen, Belgium

Vol. 1

R. S. NELSON: THE OBSERVATION OF ATOMIC COLLISIONS
IN CRYSTALLINE SOLIDS

Vol. 2

K. M. BOWKETT AND D. A. SMITH: FIELD-ION MICROSCOPY

NORTH-HOLLAND PUBLISHING COMPANY-AMSTERDAM·LONDON

# FIELD-ION MICROSCOPY

## K. M. BOWKETT
University of Cambridge Metallurgy Department
Fellow of Christ's College, Cambridge
and
## D. A. SMITH
University of Oxford Metallurgy Department
Fellow of Trinity College, Oxford

1970

NORTH-HOLLAND PUBLISHING COMPANY-AMSTERDAM·LONDON
AMERICAN ELSEVIER PUBLISHING COMPANY, INC. - NEW YORK

Publishers:

North-Holland Publishing Company – Amsterdam
North-Holland Publishing Company, Ltd. – London

Sole distributors for the U.S.A. and Canada:

American Elsevier Publishing Company, Inc.
52 Vanderbilt Avenue
New York, N.Y. 10017

Library of Congress Catalog Card Nr 76–108283
ISBN North-Holland 7204 1752 X
ISBN American Elsevier 0444 10004 0

Printed in The Netherlands

# PREFACE

*The concept of the atom as an elemental particle is thousands of years old, and the idea of an instrument with which one could* look at *individual atoms would have stirred the imagination of natural philosophers from Leucippus and Democritus to Dalton. Yet such an instrument, the field-ion microscope, now not only exists but is widely used in laboratories throughout the world to study the atomic structure of materials. In this monograph we are concerned not so much with the instrument as with its application, although since the two are so intimately related, this is more a point of emphasis than of content. The aim of the book is to indicate the applications of the field-ion microscope, both potential and realised, together with the limitations of the technique. Perhaps it will serve as a guide to those who are considering using the field-ion microscope and a reference book for those who already use it as an established research tool. It is hoped that the introductory chapters will prove to be an adequate summary of the basic processes in field-ion microscopy. It is not possible nor even perhaps desirable for a review of such a fast-growing technique to be exhaustive. Certainly we have been selective in our choice of material. Furthermore since this book is one of a series of monographs on defects in solids we have naturally tended to concentrate on this aspect of the application of field-ion microscopy. Throughout the book we have preferred, in general, to adopt the style of including the references in the text and hope that this will not prove too distracting to the reader, but we feel that this direct linking of names with specific areas of research may prove to be helpful in following up future developments.*

*In a work of this nature we have necessarily drawn heavily upon the published and unpublished work of our friends and colleagues; we are also grateful to them for active help and encouragement in many ways. It is not possible to list everyone who has helped us, but we should particularly like to mention the following to whom we are especially indebted: D. W. Bassett, D. G. Brandon, G. K. L. Cranstoun, M. A. Fortes, L. Gillott, A. Kelly, E. W. Müller, P. H. Pumphrey, B. Ralph, J. Reich, G. D. W. Smith, Miss P. J. C. Smith, Mrs. V. Smith, M. J. Southon, H. N. Southworth and P. J. Turner. In addition we are happy to acknowledge the general interest and encouragement of Professor P. B. Hirsch and Professor R. W. K. Honeycombe during the preparation of this book. We are also most grateful to all those who have allowed us to quote from their unpublished work or who have provided illustrations.*

*K.M.B.*
*D.A.S.*

*Permission to reproduce published figures has been granted by the Editors and Publishers of Acta Metallurgica (Fig. 6.4), Arkiv für Fysik (1.7, 3.3, 3.4, A2.2, A2.3), Journal of Applied Physics (2.4, 6.10), British Journal of Applied Physics (2.8, 2.11), Czechoslovak Academy of Sciences (4.8), Discussions of the Faraday Society (8.13), Fourth European Regional Conference on Electron Microscopy (7.14), Journal of the Iron and Steel Institute (6.14), Journal of the Physics and Chemistry of Solids (A1.1), Journal of Scientific Instruments (3.9, A2.1, A2.4), Metal Science Journal (5.21), Philosophical Magazine (1.4, 1.5, 1.16, 3.4(a), 3.5, 3.6, 3.7, 5.1, 5.2, 5.3, 5.5, 5.6, 5.7, 5.8, 5.9, 5.10, 5.12, 5.14, 5.15, 5.16, 5.17, 5.18, 5.19, 5.20, 6.7, 7.1, 7.2, 7.3, 7.4, 7.5, 7.6, 7.7, 7.8, 7.9, 8.5, 8.7, 8.8, 9.1, 9.5, A1.4), Physica Status Solidi (5.21), Plenum Press (2.7), Journal of the Royal Microscopical Society (9.2), Proceedings of the Royal Society (4.1, 4.2, 4.3, 4.9, 8.6), Review of Scientific Instruments (2.15), Science (3.10, 7.13), Scientific American (6.1), Surface Science (8.12, A1.3), and Springer Verlag (Berlin) (1.2).*

# CONTENTS

# 1 | PRINCIPLES OF FIELD-ION MICROSCOPY

## 1.1. Introduction

Direct observation of the atomic structure of surfaces is possible with the field-ion microscope because of its high magnification and resolution. This is not easy to achieve by other complementary techniques such as routine transmission electron microscopy. Up to the present time, however, most applications of field-ion microscopy have been in the study of extensive features such as the structure of grain boundaries, the clustering of solute atoms, interstitials or vacancies, and the morphology of precipitates. In general, the structure of these defects is still too fine to be resolved clearly by electron microscopy although it is in the area where parallel work with field-ion microscopy and electron microscopy is useful.

The investigation of features on the atomic scale, e.g. the atom positions at an interface or a dislocation core, potentially is more rewarding but, inevitably, more difficult. In this chapter the physical principles of the microscope are discussed together with methods of crystallographic analysis of field-ion micrographs. Some reviews concentrating on the physical processes are those of Müller (1960), Gomer (1961), Brandon (1966a), Southon (1966) and Turner (1967).

## 1.2. Historical development

The field-ion microscope, like its predecessor the field-emission microscope, was the invention of one exceptional man – Erwin Müller. Over a span of more than thirty years many of the major advances in both techniques have come from Müller's quite unparalleled work.

The resolution of atomic structure with the field-ion microscope was reported by Müller in 1955. However the quality of the early micrographs was poor and it was not until 1956, when the image quality had been much improved, that general interest was aroused. In 1958 at the International Electron Microscopy Conference in Berlin, Müller (1958) showed a series of micrographs of a number of materials in which the atomic structure was clearly resolved together with a number of defect configurations. Nevertheless, the technique has been adopted only slowly by other workers perhaps largely because of the early problems of specimen preparation, image interpretation, and maintainance of an adequate vacuum. General reviews are those by Brandon (1963), Müller (1965), and the proceedings of the Florida 'short course' published in 1968.

### 1.3. Principle of image formation

A field-ion microscope is based on the simple arrangement shown in fig. 1.1. The specimen is a wire that has been polished to a sharp point, 100–1000 Å in radius. It is mounted on leads which serve the dual purpose

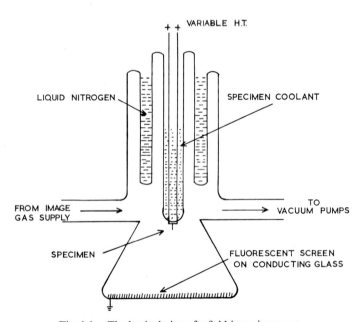

Fig. 1.1. The basic design of a field-ion microscope.

of electrical connection to the high tension power-supply and thermal contact with the coolant through which they pass. The microscope chamber is evacuated to a background pressure of less than $10^{-6}$ Torr, and a gas such as helium is leaked into the chamber to a pressure of up to $5 \times 10^{-3}$ Torr ($5\mu$). When the electric field applied to the specimen is above a critical value the gas is ionized near the specimen surface and the ions are projected on initially radial trajectories towards the fluorescent screen. Most ions leave the specimen tip from above the points of highest field which, in practice, occur above protruding atoms (fig. 1.2), so that a pencil beam of ions

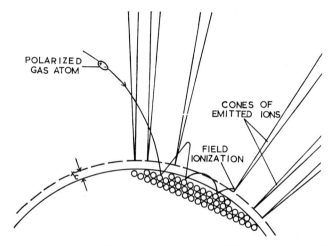

Fig. 1.2.   The image formation process: ionization of the image gas over the high field regions of a specimen tip.

travels from each such atom to excite a spot of light on the screen. Hence the image contains a bright spot for each prominent site on the specimen surface.

It is found from electron microscopy that field-ion specimens are charac-terised, after normal study in the field-ion microscope, by a smoothly curved tip. To a first approximation the tip is a spherical cap. It follows that a stack of lattice planes intersects this cap as a series of concentric rings. The rings are the loci of protruding atoms i.e. those atoms which contribute to the image by acting as centres for field-ionization. It is a reasonable approximation to regard the field-ion image as a stereographic projection of the surface field modulations. The ring systems which are small circles project therefore as small circles. In this way the familiar ring pattern of field-ion micrographs e.g. fig. 1.14 can be understood; the deviations from an exact circular form can be explained qualitatively since:

(i) the tip is not a spherical cap but a complicated curve with curvature which commonly varies by a factor two; accordingly the intersection of a plane with the surface is not a circle.

(ii) the discrete atomic structure of the specimen means that a curve is composed of a number of straightline segments; a ring is in fact a polygon bounded by low index rows of atoms.

(iii) the ion trajectories (which are not straight lines but depend on the local field) would only coincidentally result in an image which was a standard crystallographic projection.

The principle of the geometrical argument is well illustrated by the construction of a ball model of a tip bounded by a spherical surface. If the protruding 'atoms' are highlighted, a convincing model of a field-ion image results (fig. 1.3) thus justifying the previous geometrical postulates. The

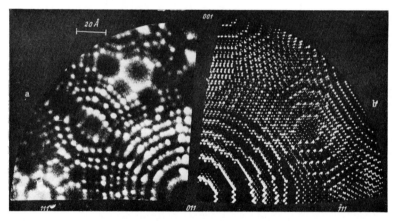

Fig. 1.3.   A ball model of one quadrant of a field-ion tip illuminated so that protruding atoms are highlighted. A corresponding quadrant of a field-ion tip is included for comparison. (Courtesy M. Drechsler and P. Wolf.)

average magnification, $M$, can be estimated to an order of magnitude by assuming a gnomonic projection which leads to the expression:

$$M = R_s/R_t \qquad (1.1)$$

where $R_s$ is the tip to screen distance and $R_t$ the average tip radius. Typical values are $R_s = 10$ cm, $R_t = 10^{-6}$ cm and $M = 10^7$.

Provided that the gas pressure is not so high that scattering of the image gas ions by collision with neutral atoms becomes important, the resolution of the image depends principally on the size of the image gas atom or molecule and its lateral velocity component; the amplitude of the thermal vibration of the specimen atoms is of lesser importance. Since the gas atoms accommodate to the tip temperature before ionization it follows that the resolution shows a marked temperature dependence; the highest specimen temperature normally used corresponds to cooling with liquid nitrogen (boiling point, 78 °K).

It is possible to remove atoms from the specimen surface by *field-evaporation*. If the atoms of the specimen evaporate at a lower field than that necessary for ionization of the image gas it is not possible to obtain a stable image, unless the imaging field of the gas can be lowered by reducing the operating temperature or by changing to an image gas of lower ionization potential.

### 1.4. Field evaporation

The preparation of a clean and smooth surface takes place in two stages. A specimen wire is polished to a fine pencil-like end form. Details of polishing conditions are given in Appendix 3. The final smoothing of the specimen surface and removal of oxide etc. are achieved by the field evaporation process. Metal ions can be evaporated from the surface if the applied field is increased to a sufficiently high value; the field, $F_E$, at which this occurs is very critical\*. Despite its fundamental importance the field evaporation process is not well understood but rapid advances in the theory may be expected following the mass spectrometric work now in progress. Field evaporation has been treated as a desorption phenomenon by Müller (1960), Gomer (1961), and Gomer and Swanson (1963). In the absence of an applied field the desorption energy $Q_0$ which is required by an atom in order to leave the specimen as a positive ion is:

$$Q_0 = \Lambda + I_n - n\varphi \tag{1.2}$$

where $\Lambda$ is the sublimation energy, $I_n$ the sum of the ionization potentials to a charge $-ne$, where $e$ is the electronic charge, and $\varphi$ the work function, which is the energy released when an electron is returned to the metal (from infinity strictly). When an ion is sufficiently far from the specimen

---

\* Since the field is altered by varying the positive voltage applied to the specimen, a voltage is sometimes specified, rather than a field e.g. the 'field evaporation voltage' is quoted for a particular specimen, rather than $F_E$.

surface for the repulsion between ion cores to be neglected, its potential in one dimension can be written as follows:

$$V(x) = -\left(nFex + \frac{n^2 e^2}{4x}\right).$$ (1.3)

Differentiation of eq. (1.3) leads to expressions for the location and magnitude of the potential energy maximum:

$$x_{max} = (ne/4F)^{\frac{1}{2}}$$ (1.4)

and

$$V(x_{max}) = -(n^3 e^3 F)^{\frac{1}{2}}.$$ (1.5)

The maximum in $V(x)$ is the 'Schottky hump'. The energy barrier to be overcome is therefore:

$$Q = Q_0 - (n^3 e^3 F)^{\frac{1}{2}}$$ (1.6)

as illustrated in fig. 1.4, Müller (1956) assumed that the barrier was overcome

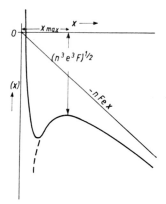

Fig. 1.4.   Schematic diagram of the energy hump to be overcome by an ion before field evaporation can occur.(Courtesy D. G. Brandon.)

by thermal activation. Gomer and Swanson (1963) and Brandon (1963 and 1966a) have pointed out that $Q$ is of the order 0.1 eV and the width of the barrier about 1 Å; under these circumstances tunnelling through the barrier should be important in low temperature field evaporation and the barrier can be regarded as partially transparent. In deriving eq. (1.6) it is assumed that the maximum in the potential energy coincides with the Schottky hump, and that the transition to the ionic state occurs before field evaporation as shown in fig. 1.5. Gomer and Swanson (1963) found the magnitude of the energy maximum by considering it to be given by the point of intersection of the atomic and ionic potential curves (fig. 1.6). In this situation the activation energy for field evaporation becomes:

$$Q = Q_0 - nFex_1 \tag{1.7}$$

where $x_1$ is the distance of the point of intersection from the specimen surface. Naturally $x_1$ must be a decreasing function of $F$ and can be determined from the atomic and ionic potential curves.

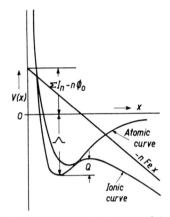

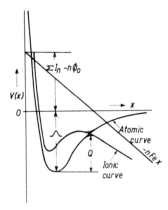

Fig. 1.5. Schematic diagram of the form of the potential energy curves when field evaporation is preceded by ionization. (Courtesy D. G. Brandon.)

Fig. 1.6. Schematic diagram of the form of the potential energy curves when ionization occurs at the moment of field evaporation. (Courtesy D. G. Brandon).

Taking field evaporation to be a thermally activated process the evaporation rate $k_e$ is given by an Arrhenius equation:

$$k_e = v \exp(-Q/kT). \tag{1.8}$$

Combining equations (1.6) and (1.8) gives:

$$F_E = n^{-3}e^{-3}[A + I_n - n\varphi - kT \ln(v/k_e)]^2. \tag{1.9}$$

As another mechanism for field evaporation Brandon (1963) considered the possibility where field evaporation occurs when the ionic potential curve intersects the potential curve of the atom at its minimum, $x_2$. The evaporation field is then given by:

$$F_E nex_2 = Q_0 - \left(\frac{n^2 e^2}{4x_2} - \frac{A}{x_2^n}\right) \tag{1.10}$$

where the term $A/x_2^n$ takes account of ion core repulsion and was assumed by Brandon to have the same form for both the atom and the field evaporated ion. By taking $x_2$ to be independent of $F_E$ a linear relation between $F_E$ and $Q$ is obtained.

Gomer and Swanson (1963) attempted a quantum mechanical treatment of field evaporation by considering the transitions from the atomic to the ionic state. In particular they showed that tunnelling could become an important feature of the field evaporation process when:

$$\left(\frac{2m}{\hbar}\right)^{\frac{1}{2}} (kT)^{\frac{3}{4}} \left(\frac{1}{S_1} - \frac{1}{S_2}\right) \leqslant 1 \tag{1.11}$$

where $m$ is the mass of the tunnelling species and $S_1$ and $S_2$ are the gradients of the potential on either side of the barrier.

Brandon (1965) has used an equation of the form (1.9) to calculate $F_E$ for many elements. It was predicted that most materials would field evaporate as double charged ions and the values of $F_E$ were in reasonable agreement with experimental data. Mass spectrometric analyses (Müller et al., 1968; Brenner and McKinney, 1968; Turner and Southon, 1969) have shown the presence of trebly and even quadruply charged ions in the evaporation of tungsten rather than a predominance of $W^{2+}$ as apparently predicted by Brandon. However Brandon noted that his calculated evaporation field for $W^{2+}$ [3.7 V/Å later (1966) revised to 5.7 V/Å] was far too low, because of the unreliability of the ionization potential data and predicted that tungsten might well evaporate as $W^{3+}$ or possibly $W^{4+}$. Turner and Southon (1969) have found the following ions in evaporation spectra of gold, iridium and molybdenum: $Au^+$, $Au^{2+}$, $Au^{3+}$; $Ir^{3+}$, $IrO_2^+$; $Mo^{3+}$, $Mo^{2+}$, and complex ions, and feel that the charge and nature of the evaporating species are sensitive to the presence of impurities. Barofsky and Müller (1968) found a, so far unexplained, temperature dependence of the ion abundances field evaporated from beryllium. Southworth (1968) has recalculated Brandon's values of $F_E$ and included corrections to take account of the different polarizabilities of the surface atom and the evaporated ion. In many cases different values are predicted for $F_E$ and sometimes the predicted charge on the evaporated ions is also different.

Within the accuracy of the model the evaporation fields at $0°K$ fall in the range 4–5 V/Å for most refractory metals and in the range 3–4 V/Å for the remaining transition metals. Since the threshold field for ionization of helium is about 3.5 V/Å at $78°K$ and may be as much as 40% lower at

TABLE   1.1

Evaporation fields calculated for some common ions by Southworth (1969, private communication) using a method devised by Brandon (1966a)

| Element | $F_1$(V/Å) | $F_2$(V/Å) | Ion expected | Element | $F_1$(V/Å) | $F_2$(V/Å) | Ion expected |
|---------|-----------|-----------|--------------|---------|-----------|-----------|--------------|
| Be | 5.4 | 4.6 | $Be^{2+}$ | Mo | 6.5 | 4.0 | $Mo^{2+}$ |
| B | 6.4 | 7.9 | $B^+$ | Ru | 6.3 | 4.1 | $Ru^{2+}$ |
| C | 14.2 | 10.3 | $C^{2+}$ | Rh | 4.9 | 4.1 | $Rh^{2+}$ |
| Mg | 2.0 | 2.5 | $Mg^+$ | Pd | 3.7 | 4.1 | $Pd^+$ |
| Al | 1.8 | 3.4 | $Al^+$ | Ag | 2.3 | 4.4 | $Ag^+$ |
| Si | 4.5 | 3.3 | $Si^{2+}$ | Cd | 2.6 | 3.1 | $Cd^+$ |
| Ca | 1.6 | 1.6 | $Ca^{2+}$ | Sn | 2.6 | 2.3 | $Sn^{2+}$ |
| Ti | 3.9 | 2.5 | $Ti^{2+}$ | La | 3.1 | 1.9 | $La^{2+}$ |
| V | 4.0 | 2.8 | $V^{2+}$ | Ta | 9.6 | 4.9 | $Ta^{2+}$ |
| Cr | 2.9 | 3.0 | $Cr^+$ | W | 10.4 | 5.9 | $W^{2+}$ |
| Mn | 2.8 | 2.8 | $Mn^{2+}$ | Re | 8.2 | 4.3 | $Re^{2+}$ |
| Fe | 4.2 | 3.4 | $Fe^{2+}$ | Os | 10.8 | 4.4 | $Os^{2+}$ |
| Co | 4.3 | 3.7 | $Co^{2+}$ | Ir | 8.5 | 4.8 | $Ir^{2+}$ |
| Ni | 3.5 | 3.5 | $Ni^{2+}$ | Pt | 6.1 | 4.4 | $Pt^{2+}$ |
| Cu | 3.1 | 4.3 | $Cu^+$ | Au | 4.8 | 5.0 | $Au^+$ |
| Zn | 2.9 | 3.5 | $Zn^+$ | Hg | 2.9 | 3.7 | $Hg^+$ |
| Ga | 1.6 | 3.9 | $Ga^+$ | Tl | 1.1 | 3.6 | $Tl^+$ |
| Ge | 3.5 | 2.9 | $Ge^{2+}$ | Pb | 2.0 | 2.3 | $Pb^+$ |
| Zr | 5.7 | 2.8 | $Zr^{2+}$ | Bi | (3.0) | 2.6 | $Bi^{2+}$ |
| Nb | 7.0 | 3.7 | $Nb^{2+}$ | | | | |

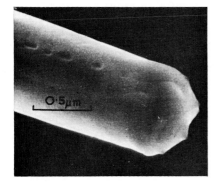

Fig. 1.7.   Scanning electron micrograph of a field-ion specimen of tungsten that has fractured under the influence of the imaging field.

5°K (Feldman and Gomer, 1963) the theory predicts that helium ion microscopy is applicable to most of the transition metals. In practice the best image field is found to be 4.5 V/Å or higher (fig. 2.7a). The rate at which field evaporation proceeds depends strongly on how far above $F_E$ the field is raised (Brandon, 1966b), but current theory overestimates the field dependence. At $F_E$ the lattice is strained to near its theoretical strength and raising the field too rapidly or too far above $F_E$ often causes the specimen to yield or fracture: the resulting surface is usually too irregular and too blunt to be of further use (fig. 1.7).

The theory developed so far is based on one dimensional arguments and contains no terms which take account of the atomic and electrical structure of the specimen surface; furthermore, the precise meaning to be attached to the distance "$x$" from the surface is not clear. Certainly, in metals the field is to be regarded as *penetrating* the surface. Gomer and Swanson (1963) add a screening term $q^{-1}$ to $x$ such that:

$$q^2 = 4m_e e^2 \left(\frac{3e_0}{\pi}\right)^{\frac{1}{2}} \hbar^{-2} \tag{1.12}$$

where $m_e$ is the effective mass of an electron and $e_0$ is the effective number of electrons per unit volume in the field-free region of the specimen.

Although atomic and ionic polarizabilities are not in general well known it is possible to write a polarization correction $\Delta Q_p$ to $Q$ as follows:

$$\Delta Q_p = \tfrac{1}{2}(\alpha_a - \alpha_i) F^2 \tag{1.13}$$

where $\alpha_a$ and $\alpha_i$ are the atomic and ionic polarizabilities respectively. $\Delta Q_p$ is usually taken to be small compared with the other terms in equation (1.9) and is often omitted. However, Müller (1964) suggests that polarization can be an important factor governing the details of the field evaporated end form. A striking case is that of the zone decorating atoms, found in images of many materials and discussed in § 3.2.4.

The polarization correction is a term which depends on the local structure of the specimen surface. Any parameter which shows a systematic orientation variation affects the end-form e.g. the structure of the surface electrical double layer and local geometrical effects such as interplanar spacings and interatomic spacings. It is to be expected that where $Q$ is lowered by the operation of the above factors the local radius will increase due to preferential evaporation and so lower the field acting (§ 1.6) to the same magnitude as at other regions of the specimen surface. Conversely where $Q$ is increased the local radius should be smaller. Possibly the major factor is the orientation dependence of work function. This dependence is a consequence of the different dipole layer effects which were considered by Smoluchowski (1941) in terms of smoothing and spreading of the electron wave functions at the specimen surface. The physical situation is that the slow decay of electron density through a relatively close packed surface leads to a double layer with negative charge outward which accounts for the relatively higher values of work function measured in such areas. In contrast the spreading of the electron cloud between the atoms in high index planes gives a double layer with positive charge outward, implying a relatively lower work function. Brandon (1965) suggested that the effects of these orientation sensitive quantities could be taken into account by rewriting equation (1.6) as follows:

$$Q = \Lambda + I_n - n\varphi - \Delta\Lambda - \Delta I - f(F). \tag{1.14}$$

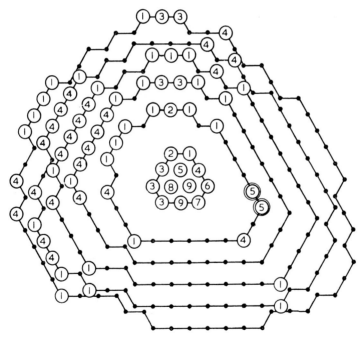

Fig. 1.8.   Diagram of the atoms in a {111} plane of a tungsten specimen; the atoms were removed during field evaporation in the sequence shown. With the exception of those indicated by a double ring all the atoms were removed in the order expected from consideration of co-ordination number. (Courtesy A. J. W. Moore and J. A. Spink.)

The activation energy $Q$ now refers to a *particular atom*. $\Delta A$ takes account of co-ordination number, position relative to lattice defects, and the species of neighbouring atoms, whilst $\Delta I$ is the difference in ionization potential of different species and is relevant to the analysis of alloy field evaporation. The correction $f(F)$ allows for differences in the effective field acting at particular sites. The field acting at a site, $F_a$, is related to the average field $F$, determined, for instance, from a smooth section approximation by the expression:

$$F_a = \beta_1 \beta_2 \beta_3 F \tag{1.15}$$

where $\beta_1$, $\beta_2$, and $\beta_3$ are enhancement factors relating to local radius, local geometry, and species of neighbours. $\beta_1$ takes account of the fact that $F$ is an inverse function of the local tip radius; $\beta_2$ reflects the increase in field at the edge of a closely packed atom plane; and $\beta_3$ allows for charge exchange between neighbours of different species (a local positive charge implies field enhancement).

Some broad generalizations can be made in terms of this model (Brandon, 1965). The low index planes have bigger step heights than high index planes and a correspondingly greater enhancement of the field at their edges. This, together with the structure of the surface electrical double layer which increases $\varphi$ for low index planes, means that the low index regions of a specimen may be expected to be of larger radius of curvature than the high index regions, a prediction that is realised in practice. The activation energy for field evaporation, $Q$, depends on the binding energy of the atom to be field evaporated which, in the perfect lattice, is a function of co-ordination number. The atoms with the lowest $Q$ are those occupying kink sites*. The density of kink sites is not constant and is lower in low index regions than in high index regions. If field evaporation is to proceed without a change in the nature of the end form it follows that $Q$ must be less in low index regions. A further restriction on $Q$ is that a stable end-form is maintained only if the rate of removal of material from the tip axis is greatest and if the rate decreases towards the shank.

The field evaporation behaviour of alloys is treated more fully in chapter 9. The crucial effect is that the presence of a second species leads to the introduction of changes in $Q$ which do not vary systematically except

---

* A surface atom on a kink site has a co-ordination number that is reduced below the normal value for a surface atom, but not reduced so far that it becomes merely an adatom. Examples of atoms in kink sites are those marked 4 in the outer "ring" of fig. 1.8.

in fully-ordered alloys. As a result the field evaporated end-form becomes irregular and the image fails to show the usual ring pattern. This effect is more marked when the solute is preferentially retained, than when it is preferentially field evaporated. The images from (disordered) concentrated solid solution alloys are less regular than those from dilute alloys.

The field evaporation process is paradoxically both the chief asset and the major limitation of field-ion microscopy. Only the most refractory materials evaporate at a field appreciably above the field necessary for field-ionization of helium, and thus give stable images, with liquid nitrogen cooling. To study less refractory materials it is necessary either to change to a gas of lower ionization potential or to reduce the temperature of the specimen and image gas, which also lowers the threshold field for ionization.

The atoms on a stable surface which can be most readily evaporated are also those being imaged. By careful control of the field evaporation process it is possible to remove only those atoms occupying the sites which have already been examined, and permit the neighbouring atom sites to be scrutinized. The manner in which this is done experimentally is to raise the voltage applied to the specimen until the onset of field evaporation and then to reduce the voltage rapidly to the *best image voltage* once a few atoms have been removed. Alternatively a small positive pulse may be superimposed on the high voltage supply to the tip thus permitting control of the field evaporation process (Brenner, 1965). By taking a series of micrographs after successive small amounts of field evaporation, making allowance for distortions and rearrangements, it is possible to build up a three dimensional picture of the specimen structure.

## 1.5. Field ionization

When the high electrostatic field is acting on the specimen, neutral gas atoms or molecules arrive at the specimen surface at a much greater rate than would be expected from simple kinetic theory, because of the polarization of gas molecules induced by the applied field. After striking the specimen surface a gas atom or molecule rebounds. Provided sufficient kinetic energy has been lost to the lattice, the polarized gas atom or molecule is unable to escape from the dipole attraction of the tip. It executes a series of jumps of decreasing amplitude shedding energy to the lattice at each impact and remaining for increasing times in the region of high ionization probability near the surface. When sufficiently accommodated to the specimen temperature, gas atoms or molecules lose an electron to the tip by tunnelling.

Ionization will not occur at less than a certain critical distance (probably about 5 Å) from the surface.

The possibility that the process which occurs might be tunnelling of atoms or ions was ruled out as a result of work by Inghram and Gomer (1954) who used a mass spectrometer to analyse the positive ions formed from various image gases in a field-ion microscope. It was found that hydrogen and deuterium formed similar ions at almost identical fields thus excluding the tunnelling mechanism which should be strongly mass dependent.

The process of electron tunnelling is considered with reference to the potential diagrams shown in fig. 1.9 (Southon, 1963). Figure 1.9(a) shows the potential of an outer electron of an atom before an external field is applied: the electron lies at a potential $I$ below the energy zero. In fig. 1.9(b) a field has been applied to the atom and ionization can occur by the electron

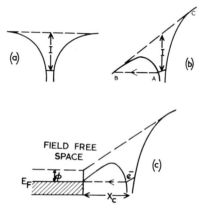

Fig. 1.9. Potential diagrams for an electron in field-ionization: (a) the potential due to the parent ion in zero field and, (b) in a field BC, (c) the potential in an applied field near the surface of a metal of work function $\varphi$. (After M. J. Southon.)

tunnelling along a path such as AB through the potential barrier. The tunnelling probability is increased for an atom near a conductor as illustrated in fig. 1.9(c) where the height of the potential barrier is reduced, although the field remains the same. This is because of the attraction between the electron and the image charge distribution induced in the metal. However, if the gas atom comes closer than the critical distance, $x_c$, shown in fig. 1.9(c), tunnelling is unlikely to occur because the energy of the electron in the gas atom will be below the Fermi energy of the metal and there is a very low density of vacant energy levels available for the tunnelling electron.

The qualitative description of field-ionization can be re-written quanti-tatively, although simple and not entirely realistic models are used. The probability of tunnelling through a potential barrier of the kind indicated in fig. 1.9(c) can be estimated from the WKB solution to the potential:

$$p = \left\{ \frac{2m_e}{\hbar^2} (E - V(x)) \right\}^{-\frac{1}{2}} \exp - \left( \frac{8m}{\hbar^2} \right)^{\frac{1}{2}} \int_{x_2}^{x_1} [V(x) - E]^{\frac{1}{2}} \, dx \qquad (1.16)$$

where $m_e$ is the electron mass, $E$ its total energy, $V(x)$ the potential of the tunnelling electron at a distance $x$ from the surface and the limits of integra-tion refer to the barrier width. Good and Müller (1956) give the following expression for $V(x)$:

$$V(x) = \frac{-e^2}{x_n - x} + Fex - \frac{e^2}{4x} + \frac{e^2}{x_n + x} \qquad (1.17)$$

where the first term is the Coulomb attraction by the ion taken to be at a distance $x_n$ from the surface, the second term is the potential due to the applied field and the last two terms are the attraction and repulsion of the electron and ion images respectively. The approximate nature of the expres-sion for $V(x)$ is clear since at both $x=0$ and $x=x_n$, $V(x)$ is unbounded. On physical grounds $V(x)$ must join smoothly to the curve describing the potential in the solid.

Alternatively, Gomer (1961) showed that the probability of barrier penetration was given by the following approximate expression in which numerical values of constants have been inserted:

$$p = \exp \left[ -0.68 \frac{I^{\frac{3}{2}}}{F} \left( 1 - \frac{7 \cdot 6 F^{\frac{1}{2}}}{I} \right)^{\frac{1}{2}} \right] \qquad (1.18)$$

where $I$ is in eV and $F$ in V/Å.

The critical distance for field-ionization, $x_c$, can be calculated to a good approximation by assuming that it is the separation of gas atom and surface at which the electron energy coincides with the Fermi energy. This condition may be written (Müller, 1960) after making an energy balance:

$$Fex_c = I - \varphi - \frac{e^2}{4x_c} + \frac{1}{2}(\alpha_A - \alpha_I) F^2 \qquad (1.19)$$

where $I$ is the ionization potential of the gas atom, $-e^2/4x_c$ is the potential due to the ion and its image, and the last term is the difference between the polarization potential energy of the gas atom and the resultant ion. Neglecting the last two terms, which are smaller than the first two, gives the following expression for $x_c$:

$$x_c = \frac{I - \varphi}{Fe}.$$  (1.20)

Substitution of appropriate values in (1.20) suggests that $x_c$ is about 5 Å for the gases likely to be used in field-ion microscopy.

The probability of ionization, $dp_i$, whilst a gas atom is travelling a distance $dx$ normal to the surface depends on $p$, $V_n$ the velocity normal to the surface and on the frequency, $v$, with which the electron approaches the barrier:

$$dp_i = \frac{dx}{V_n \tau_i(x)}$$  (1.21)

where $\tau_i(x)$ is a lifetime for ionization at a distance $x$ from the surface. $\tau_i(x)$ is defined as follows:

$$\tau_i(x) = (v p_i)^{-1}.$$

The variation of ionization probability with $x$ can then be written:

$$\frac{dp_i}{dx} = \frac{1}{V_n \tau_i(x)} = \frac{vp}{V_n}.$$  (1.22)

Experimental measurements (Southon and Brandon, 1963) show that $p$ is a sensitive function of the applied field $F$ which in turn depends strongly on $x$; $dp_i/dx$ is therefore expected to increase sharply as $x$ decreases towards $x_c$ in accord with the calculations of Müller and Bahadur (1956) and Southon (1963). A consequence of the steep increase in $dp_i/dx$ as $x_c$ is approached is that field-ionization should occur in a narrow zone just greater than $x_c$ from the surface.

The experiments of Tsong and Müller (1964) using a retarding potential energy analyser enabled the half width of the energy distribution of helium ions formed above a tungsten emitter to be determined as 0.60 eV. This small value corresponds to an ionization zone only 0.136 Å wide. Jason et al. (1966) and Jason (1967) have also analysed the energy spectra of field-ions using the tungsten-hydrogen system. A well defined oscillatory structure was found in the ion energy distribution suggesting that field-ionization also occurs at distances rather greater than $x_c$. They postulated that this arose from the production of resonant states by the action of the applied field and obtained semi-quantitative agreement between calculation and experiment. In this way, with the assumption of an infinite triangular potential well bounded by the surface and the applied field, the energies corresponding to the peaks can be deduced from a quantum mechanical treatment.

Boudreaux and Cutler (1966a and 1966b) also used a quantum mechanical formulation and applied time-dependent perturbation theory to both one and three dimensional models of the field-ionization process. Good agreement was obtained with the experimental determination of the width of the ionization zone by Tsong and Müller (1964). The probability of ionization during a single pass through the ionization zone was calculated to be of the order $10^{-9}$.

Duke and Alferieff (1967) have calculated the barrier penetration probabilities for a number of different barrier configurations including situations where the surface was partially covered by adsorbate.

Knor and Müller (1967) have proposed that field-ionization models should include consideration of the interaction between the electron orbitals of the gas atom and those of the metal surface. In this way it is suggested that anomalous details of the field-ionization process e.g. the $60°$ alternating contrast in (0001) of rhenium can be understood. The considerable variations in brightness of the various regions of the image are characteristic of the crystal structure and material of the specimen and are likely to be governed by differences in field-ionization behaviour. Moore and Brandon (1968) point out that the crystallographic variations in brightness correspond to discrete changes in co-ordination number. It is not clear why this correlation exists.

The ion current arriving at the microscope screen can be measured and plotted as a function of the voltage applied to the specimen. Figure 1.10 shows typical variations of helium ion current with voltage on a log-log plot for a tungsten specimen at various temperatures. The practical significance of the form of these curves is considered in § 2.6. The field-ion current from any ionization centre on a tip is governed by the local ionization probability and the supply of gas atoms. The curves show two regions. At low field strengths the ion current is limited by the ionization probability since the number of ions produced is small compared with the supply of neutral gas atoms. At high field strengths the ionization probability approaches unity and the ion current is governed by the supply of gas atoms. The slope of the initial part of the curves in fig. 1.10 is about 30 and is not temperature dependent. The latter part of the curve has a slope of about 3 and is temperature dependent.

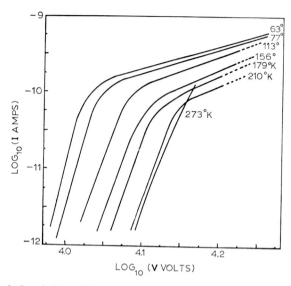

Fig. 1.10.  Typical variations of helium ion current with voltage for a tungsten specimen at various temperatures on a log-log plot. (Courtesy M. J. Southon.)

The supply of gas atoms exceeds that expected from gas kinetic theory because of the strong polarizing action of the high field acting on the specimen. Southon (1963) has evaluated the effect, and, for a gas of polarizability $\alpha$ obtains a supply enhancement factor of $(9\alpha F^2/8kT)^{\frac{1}{2}}$ over the simple kinetic theory expression. This factor is of magnitude 10–100 in practice.

Once formed, ions are projected from the specimen at the points of highest field (protruding atoms) and travel on initially radial trajectories towards the fluorescent screen. The actual trajectories are not known in detail but their form is important since it governs the nature of the projection. It will be appreciated that the field-ion image is not of the actual atoms but of the electric field distribution some small distance above the surface. This distinction does not appear greatly to affect image interpretation in practice.

## 1.6.  The field and the stress

In a field-ion microscope the specimen is subjected to very high electric fields and consequent stresses. To determine the effect of these stresses it is first necessary to know the field at the tip surface. There are two analytic methods for solving this problem:

1) to solve Laplace's equation for the region between the screen and the tip,

2) to consider a distribution of charges which gives a family of equipotentials such that one coincides with the tip surface and one with the screen.

In the first approach the main difficulty is to fit the boundary conditions. The usual method of solution is to work in a co-ordinate system, say $(\xi, \eta)$ (axial symmetry is assumed throughout so only two co-ordinates are needed) where the tip and screen surfaces are represented by $\xi =$ constant or $\eta =$ constant. Eyring et al. (1928) found a solution in this way using hyperbolic co-ordinates and a tip was assumed to be hyperboloidal in shape. Electron micrographs of field-ion tips show that in general this is not a good approximation, the shape of the tip being closer to a paraboloid. Becker (1951) and Beckey et al. (1968) used parabolic co-ordinates and assumed both the tip and the screen to be paraboloidal in shape. Because the tip to screen distance is typically of the order of $10^6$ times larger than the radius of the tip, the fact that the screen is planar should not affect the field at the surface of the tip very much. However, should the field throughout the whole region between the screen and the tip be required, e.g. for calculations of ion trajectories, then the actual shape of the screen must be considered and the paraboloidal approximation is invalid.

Following the second approach, Dyke et al. (1953) found that the geometry of a typical emitter could be fitted accurately by one equipotential surface from the family surrounding a charged body which consisted of an isolated sphere on an orthogonal cone. If it is assumed that the screen is another equipotential from the same family, the potential distribution may be written:

$$V = \frac{V_R}{R_s^n}(r^n - a^{2n+1} r^{-(n+1)}) P_n(\cos\theta) \tag{1.23}$$

where $V_R$ is the potential between the body and the screen, $r$, $\theta$, are plane polar co-ordinates measured from the sphere centre, $R_s$ is the tip to screen distance, $a$ is the sphere radius, and $P_n$ is a Legendre function with $n$ chosen so that $P_n = 0$ when the exterior half angle of the cone, $\alpha$, equals $\theta$. The screen is assumed to be at zero potential, $V_{\text{screen}} = V_R$, and if the term $a^{2n+1} r^{-(n+1)}$ is neglected the equation of the screen can be written as $r^n = R_s^n/P_n (\cos\theta)$. This function describes an approximately paraboloidal surface, and again the solution is not applicable to calculations of ion trajectories.

In order to compare their theory with experimental results Dyke et al.

defined a quantity $\beta = F/(V_R - V)$ at several points on the tip, where $F$ is the field at the required point and $V$ is the potential on the surface substituted for the emitter. From these values the emitting area was predicted and was found to be in good agreement with experimental results, which confirms that eq. (1.23) is valid for a *curved* screen.

Fortes (1968) considered a semi-infinite string $z \geqslant 0$ ($z$ axis is the axis of the tip) along which there was a uniform line density of electric charge $\lambda > 0$ plus a charge $q$ at the origin of co-ordinates. In spherical co-ordinates $\rho$ (distance from origin), $\theta$ (measured from $z$ axis) the field at any point is:

$$F = \frac{\lambda^2}{\rho^2 \sin^2 \tfrac{1}{2}\theta} + \frac{q^2}{\rho^4} + \frac{2q\lambda}{\rho^2} \tag{1.24}$$

$q$ being determined so that an equipotential, $V =$ constant, has the approximate shape of a field-ion tip. When $-q$ is put equal to $R_s \alpha \lambda$, it is found that $\alpha = 0.4$ gives the most uniform field distribution on the surface, although the field still varies by a factor of 1.4 for $\theta$ between $0°$ and $52°$. Fortes also calculated:

$$\frac{V}{FR_t} = 0.64 \ln\left(\frac{R_s}{R_t}\right) + 0.39$$

where $R_s =$ screen to tip distance. Using typical values, tip radius, $R_t = 500$ Å, $R_s = 5$ cm he found $V/FR = 4.13$ (cf. Gomer's (1961) empirical value of 5). Tayler and P. J. Smith (unpublished) have obtained a more detailed solution for the electric field by a combination of the two methods described above. They considered first a distribution of charge $\rho(s)$ per unit length along the tip axis. The potential at any point $(z, r)$ is then:

$$V(z, r) = \int_0^L \rho(s) \left\{ [(z - s)^2 + r^2]^{\frac{1}{2}} - [(z + 2R_s + s)^2 + r^2]^{\frac{1}{2}} \right\} \, ds \tag{1.25}$$

where $(z, r)$ are cylindrical co-ordinates, $z$ being the tip axis and the origin being at the tip vertex. $R_s$ is the tip to screen distance and $L$ is the length of the specimen. When the slender body theory developed for hydrodynamics is used it is found that $\rho(s) = -V/2 \ln R_t(z)$ where $r = R_t(z)$ is the equation of the specimen. If it is assumed that $L/R_s \ll 1$ and $R_t(L)/L \ll 1$ the potential near the body, but at a distance greater than $\varepsilon$ from the vertex, where $\varepsilon$ is the radius of curvature at the vertex, is:

$$V = V_0 \frac{\ln r}{\ln R_t(z)} + \text{smaller terms}. \tag{1.26}$$

This gives the field on the body as:

$$F = -\frac{V_0}{R_t(z)\ln R_t(z)\cos\beta} \qquad (1.27)$$

where $\beta$ is the angle between the local normal to the surface and $r$.

In order to obtain a solution near the vertex Laplace's equation was solved by first approximating the vertex region by a paraboloid and using paraboloidal co-ordinates defined by

$$z + ir = d - c(\xi + i\eta)^2$$

where $c$ and $d$ are constants. This gives:

$$V(\xi) = V_0 \frac{\ln a\xi}{\ln a\varepsilon} \qquad (1.28)$$

where $a$ is a constant of integration. Since $a$ is arbitrary the solution is indeterminate, therefore the process was repeated in ellipsoidal co-ordinates with the tip approximated by an ellipsoid, which gave $a = \frac{1}{2}$, i.e.

$$V = V_0 \frac{\ln \frac{1}{2}\xi}{\ln \frac{1}{2}\varepsilon}. \qquad (1.29)$$

With this expression transformed back to cylindrical co-ordinates, and the tip fitted by an ellipse of semi-major axis $g$, the field was found to be:

$$F = -\frac{2V_0}{[R_t(z)^2 + \varepsilon^2]^{\frac{1}{4}}\ln \varepsilon/4g}. \qquad (1.30)$$

When this expression was applied to data from electron micrographs of several typical tips the ratio of the field at $60°$ from the axis to the field at the vertex was found to be about $1:1.05$, which is in good agreement with experimental field-ion observations. One application of expressions such as the above is the estimation of the stress acting in a field-ion specimen.

Rendulic and Müller (1967) have considered the expansion of a field-ion tip under the negative pressure caused by the field. Uniform expansion would not be observable on a field-ion micrograph. However, non-uniform expansion occurs because of the varying field strength over the tip and the variation of the elastic constants of the material with crystallographic direction. The differences in projection angle between two particular atoms in a tungsten surface imaged at varying voltages using different image gases were measured and plotted against (voltage)$^2$. The plot was linear and, since the stress $\sigma$ caused by a field $F$ is $\sigma = F^2/8\pi$ it was concluded that the

deformation is proportional to the stress, at least to a first order approximation. The measured displacements were reproducible after two hours so it
was concluded that they were elastic. Using eq. (1.30) to find the field
P. J. Smith and D. A. Smith (1970) have shown that the average tensile force
on a plane surface through the tip, perpendicular to the axis at $z = z_0$ is given
by:

$$F_z = \frac{V_0^2}{2\pi} \frac{\ln\left[1 + (r_0/\varepsilon)^2\right]}{[\ln \varepsilon/4g]^2}. \tag{1.31}$$

At $z_0 = 100$ Å, this force produces a stress of about $7 \times 10^{10}$ dynes/cm$^2$.
When $z_0 \gg \varepsilon$, the tensile force is calculated by eq. (1.27) and eq. (1.30), and is
given by:

$$F_z = \frac{V_0^2}{4 \cos^2 \beta}\left[\frac{1}{\ln r_1} - \frac{1}{\ln r_0}\right] + \frac{V_0^2}{2}\frac{\ln\left[1 + (r_1/\varepsilon)^2\right]}{[\ln \varepsilon/4g]^2} \tag{1.32}$$

where $r_1 = R(z_1)$, $z_1$ being the value of $z$ at which eq. (1.27) becomes valid.
Under typical imaging conditions, this force produces a stress of the order
of $5 \times 10^{10}$ dynes/cm$^2$ at a point about 5000 Å from the vertex of the tip,
fig. 1.11. The stress in the shank and the stress near the tip are of the order
respectively of the technical and theoretical fracture stresses of the refractory
metals. It might be expected, therefore, that field-ion specimens could fracture
either by a conventional failure in the shank or a fracture process at the tip
near the theoretical stress. Failure of the *tip* at the technical fracture stress
does not generally occur (fortunately, or field-ion microscopy would not be
possible) because dislocation sources are absent. However, dislocation
loops produced by homogeneous nucleation have been observed in iridium
by Fortes et al. (1968) and Fortes and Ralph (1968) and analysed by Smith
and Bowkett (1968) in tungsten specimens. Conventional yield behaviour
initiated in the shank is quite common. In the bcc metals yielding often
leads to fracture; in the low modulus fcc and hcp materials such as platinum,
gold, and cobalt slip bands which intersect the tip surface are a frequent
artefact. Yielding can be minimised by operating with low specimen temperatures; this technique is most effective with the bcc metals which show
a strong dependence of the yield stress on temperature. The complete
condition for stable field-ion images to be obtained becomes:

$$F_Y > F_E > F_I \tag{1.33}$$

where $F_Y$ is the field at which a yielding process occurs.

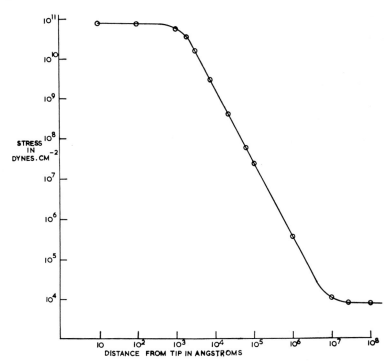

Fig. 1.11.   Log-log plot of the tensile stress in a field-ion specimen during helium imaging as a function of distance along the axis from the tip. The apex radius ($\varepsilon$) is about 1000 Å.

## 1.7. Analysis of field-ion micrographs

Drechsler and Wolf (1958), in a classic paper, showed how to deduce the Miller indices of a pole in a field-ion micrograph, the local tip radius and the local magnification. They rea ised that the prominence of a particular pole depends directly on the inter-planar spacing. (In multiply primitive lattices such as fcc and bcc it is important to state the spacing of the atom planes correctly, e.g., in bcc if $h+k+l$ is not even then $2h$, $2k$ and $2l$ must be substituted in the formula for interplanar spacing; similarly in fcc, if $h$, $k$, and $l$ are mixed, $2h$, $2k$ and $2l$ should be substituted in the formula.) Drechsler and Wolf's analysis was applied only to field-ion micrographs of tungsten.

The measurements of Brandon (1964b) on a micrograph of tungsten suggested that the field-ion projection corresponded to one of the same family as the gnomonic and stereographic projections but with the point of projection situated one radius outside the projection sphere as shown in fig. 1.12. Fortes (1968) from his analysis of field-ion micrographs of iridium suggested that it was a useful device to regard the projection point as being in a different position for the various crystallographic regions of the image which were then individually closer to being stereographic projections. Adopting an empirical approach, having realised that the field-ion projection was not simple, Newman, Sanwald and Hren (1967) applied a generally applicable relation of the form:

$$l = MR\theta^1. \tag{1.34}$$

Equation (1.34) states the relation between the linear separation $l$ of two poles inclined at an angle $\theta$ where a projection sphere has a radius $R$, and $M$ is an empirically determined factor. This geometry is not illustrated in fig. 1.12. Such a relation is very similar in practice to that proposed by Brandon (1964b) and will be recognised as the first term of the Taylor expansion of a general projection relationship $l=f(R, \theta)$.

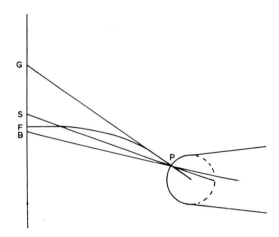

Fig. 1.12.  Some possible projections for the field-ion image. F marks the approximate position where an ion beam from a point strikes the screen; F can be compared with S, G, and B which are projections of the same point using the stereographic, gnomonic, and Brandon (1964) projections.

Figure 1.13 shows electron micrographs of the field evaporated end forms of tungsten, iridium and molybdenum. It can be seen that all the surfaces are smoothly curved but the details of the profile are peculiar to the material concerned and cannot be described by a simple analytical function.

Figure 1.12 compares some possible projections. That the field-ion projection is not a gnomonic projection follows from the observation that the solid angle subtended at the field-ion tip by the screen is always substantially less than the crystallographic angle imaged on the screen. In reality, a field-ion micrograph cannot be regarded as being any particular projection. It is evident from fig. 1.13 that the local radii of the tips vary from point to point. The projection is expected to vary over the image, to depend on the material, and to differ from microscope to microscope.

All three standard projections, gnomonic, stereographic and orthographic (and indeed any axial projection), allow direct measurement of the angle between zones intersecting at the centre of the projection. This property holds for the field-ion projection too, although the central pole does not necessarily exactly define the centre of the projection. Symmetry elements can be recognised readily and some are marked in fig. 1.14 which is a tungsten field-ion micrograph. Having identified and indexed the symmetry elements apparent, it is then comparatively straightforward to index other poles using the zone rule and the relative prominence rule. As shown in fig. 1.14 high index regions can be identified in this way. Since in planes such as the top layer of the $(25\bar{1})$ pole every atom is resolved, it is possible to verify the indexing directly by comparing observed and expected atom arrangements. Such a procedure can be justified only in poles inclined at small angles to the tip axis and for closely spaced image spots in which case the distortions are not too great. Figure 1.15 shows the detailed crystallographic arrangement of the four atoms constituting the top layer of the $(25\bar{1})$ pole of fig. 1.14. The correspondence between the image and the model is immediately apparent. Examples of situations where the field-ion image does not reflect the atom arrangement are discussed later in this section.

Table 1.2 summarizes measurements of angle distance relationships made on the tungsten specimen shown in fig. 1.14 and on iridium and molybdenum specimens photographed under identical conditions. It can be seen that the stereographic projection is a reasonable approximation for small angles but subject to considerable errors where large angles are involved. The details of the deviation depend on the specimen material.

The assumption of a stereographic projection must sometimes be made without detailed justification in order to simplify interpretation. However, it is necessary to be cautious about quantitative conclusions based on the stereographic assumption.

Knowing the atom arrangement in a particular fully resolved pole it may be possible to determine the magnification directly by measuring the

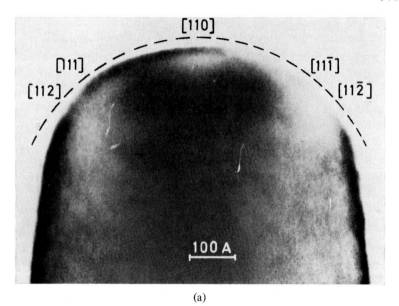

(a)

(Courtesy B. Loberg and H. Nordén.)

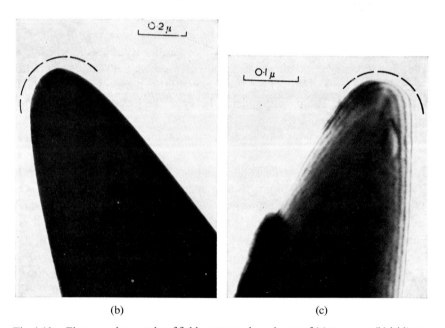

(b)                                          (c)

Fig. 1.13.   Electron micrographs of field evaporated specimens of (a) tungsten, (b) iridium, and (c) molybdenum. Note how the local radius varies from point to point.

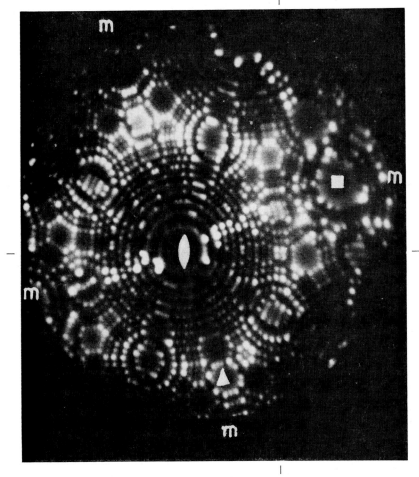

Fig. 1.14. Some symmetry elements marked on a tungsten micrograph in the (110) orientation. The lines mark the $x$ and $y$ co-ordinates of the (25$\bar{1}$) pole.

atom separation on the micrograph. There is an uncertainty in the magnification deduced in this way because of the unquantified effects of the surface and the imaging field, and the dilatation in a field-ion tip during helium imaging. This can be estimated to be several percent (by an extension of the arguments in § 1.6), depending on the material and imaging conditions*.

* The dilatation in tungsten during helium imaging under a hydrostatic stress of $3 \times 10^{11}$ dynes $cm^{-2}$ estimated by this method is about 12%; however the actual change in lattice parameter will only be from 3.16 Å to 3.28 Å (3.9%). In the case of iron imaged in neon (Vitek, private communication) the change in lattice parameter is much less: about 1%.

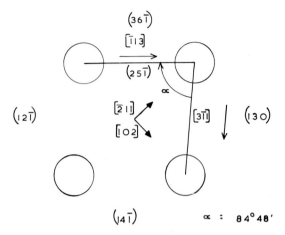

Fig. 1.15. The detailed crystallographic arrangement of the four atoms constituting the top layer of the (25$\bar{1}$) pole of fig. 1.14. The figure is unfortunately drawn with a 90° clockwise rotation relative to fig. 1.14; however, the correspondence between the image and the model is apparent.

Furthermore, good correspondence between expectation and observation of atomic positions is not always found. For instance, fig. 1.16 shows the (021) region of a tungsten field-ion tip. The atoms appear to be arranged on a square grid. However, the atom arrangement in (021) is based on a rectangular unit mesh with sides in the ratio $\sqrt{5}:1$.

The (021) pole occurs on the [100] zone which shows a striking feature in tungsten images known as *zone decoration* (§ 3.2.4). Müller (1964, 1967a) suggested that the atoms constituting the zone decoration were stabilized in protruding low co-ordination number sites by a polarization bonding contribution due to the applied field. These sites are not necessarily expected to be normal lattice sites. Indeed not all the atoms in the (021) pole always appear to be in a square array: an example of an arrangement in (021) which spuriously gave the appearance of being a classical edge dislocation has been observed by Müller (1960).

### 1.8. Some geometrical relationships

#### 1.8.1. MAGNIFICATION

A simple expression for the average magnification $M$ as a function of the tip to screen distance, $R_s$, and the average tip radius, $R_t$ is:

$$M = R_s/R_t.$$

TABLE 1.2

Comparison of measured values, $l$, of the separation of poles $hkl$ and $h'k'l'$ inclined at an angle $\alpha$, with the stereographic prediction $p$. For ease of comparison $l$ values and $p$ values have been divided by $l_{110\text{-}231}$ and $p_{110\text{-}231}$ respectively for tungsten, $l_{001\text{-}115}$ and $p_{001\text{-}115}$ respectively for iridium and $l_{110\text{-}22\bar{1}}$ and $p_{110\text{-}22\bar{1}}$ respectively for molybdenum

### Tungsten

| $hkl$–$h'k'l'$ | $\alpha$ | $l$ (measured cm) | $l/l_{110\text{-}231}$ | $p/p_{110\text{-}231}$ |
|---|---|---|---|---|
| 110–111 | 35°16′ | 6.70 | 1.76 | 1.85 |
| 110–112 | 54°44′ | 9.50 | 2.50 | 3.01 |
| 110–121 | 30° | 5.65 | 1.49 | 1.56 |
| 110–231 | 19°6′ | 3.80 | 1 | 1 |
| 110–132 | 40°54′ | 7.25 | 1.91 | 2.17 |
| 110–011 | 60° | 11.50 | 2.76 | 3.35 |
| 110–141 | 33°34′ | 6.30 | 1.66 | 1.71 |
| 110–031 | 47°54′ | 9.00 | 2.37 | 2.58 |
| 110–130 | 26°33′ | 5.10 | 1.34 | 1.37 |
| 110–010 | 45° | 8.70 | 2.29 | 2.46 |

### Iridium

| $hkl$–$h'k'l'$ | $\alpha$ | $l$ (measured cm) | $l/l_{001\text{-}115}$ | $p/p_{001\text{-}115}$ |
|---|---|---|---|---|
| 001–115 | 15°48′ | 2.65 | 1 | 1 |
| 001–113 | 25°18′ | 4.05 | 1.53 | 1.62 |
| 001–112 | 35°16′ | 5.32 | 2.01 | 2.29 |
| 001–335 | 41°49′ | 6.25 | 2.36 | 2.75 |
| 001–111 | 54°44′ | 9.00 | 3.40 | 3.73 |
| 001–102 | 26°34′ | 4.35 | 1.64 | 1.70 |
| 001–101 | 45° | 7.10 | 2.68 | 2.98 |

### Molybdenum

| $hkl$–$h'k'l'$ | $\alpha$ | $l$ (measured cm) | $l/l_{110\text{-}23\bar{1}}$ | $p/p_{110\text{-}23\bar{1}}$ |
|---|---|---|---|---|
| 110–23$\bar{1}$ | 19°6′ | 4.55 | 1 | 1 |
| 110–22$\bar{1}$ | 19°26′ | 4.10 | 0.90 | 1.02 |
| 110–33$\bar{2}$ | 25°12′ | 6.00 | 1.32 | 1.34 |
| 110–11$\bar{1}$ | 35°16′ | 8.20 | 1.81 | 1.89 |
| 110–12$\bar{1}$ | 30° | 6.95 | 1.53 | 1.59 |
| 110–010 | 45° | 9.20 | 2.02 | 2.46 |

The derivation of this equation assumes a gnomonic projection and is therefore an overestimate. Similar expressions can be derived for other projections.

Following Drechsler and Wolf (1958), the local radius, $R$, may be estimated by counting the number of rings, $n$, between two poles separated

Fig. 1.16.   The (021) region of a tungsten specimen imaged with helium at about 27°K.

by a known angle $\theta$:

$$R = \frac{n d_{hkl}}{1 - \cos \theta} \tag{1.35}$$

where $d_{hkl}$ is the appropriate step height. In fig. 1.17 the radius from $(11\bar{1})$ to $(11\bar{2})$ is clearly less than that either from $(11\bar{1})$ to $(12\bar{1})$ or $(11\bar{1})$ to $(21\bar{1})$; the latter two are about the same. The ratio of $R_{(11\bar{1})/(11\bar{2})}$ to $R_{(11\bar{1})/(12\bar{1})}$ is about 4:7 whilst the local magnifications are in the ratio 7:5 i.e., there is, as expected, an inverse relationship between local radius and local magnification. This is consistent with the well known fact that the average magnification decreases as the average radius increases with field evaporation.

If it is necessary to make frequent estimations of the radius of specimens of a particular material eq. (1.35) can be simplified to the form:

$$R = k \cdot n \; ; \tag{1.36}$$

$k$ is a parameter which depends upon the specimen material and the indices

$(21\bar{1})$

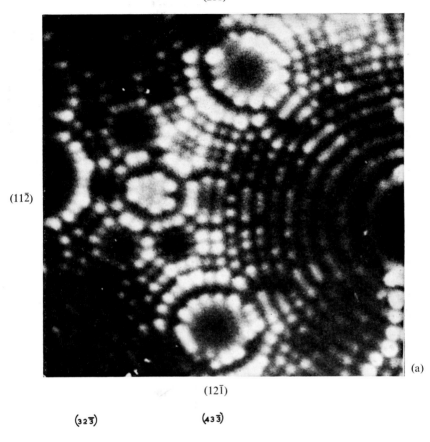

$(11\bar{2})$

(a)

$(12\bar{1})$

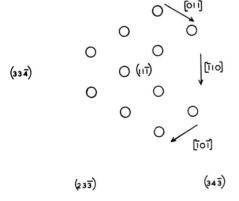

$(32\bar{3})$     $(43\bar{3})$

$(33\bar{4})$     $(33\bar{2})$

$(23\bar{3})$     $(34\bar{3})$

Fig. 1.17. (a) Selected area of a field-ion micrograph of tungsten showing the region surrounding the $(11\bar{1})$ pole $(78\,^{\circ}\mathrm{K}$; helium image gas). (b) Plan of the atom arrangement in the $(11\bar{1})$ pole of (a). Comparison of this plan with the microgrpah shows the distortion of the image.

(b)

of the field-ion poles between which the number of rings, $n$, is counted. For example, when micrographs of tungsten are being analysed it is convenient to count the rings between the (110) and (321) poles, in which case $k$ equals 33.

One result which can be predicted, in view of the nature of the projection, is an increase in magnification in poles inclined at large angles to the tip axis. Most micrographs show this effect; the apparent ledge widths of peripheral poles are greater than those of the crystallographically equivalent poles close to the centre, yet the local radii should be nearly the same.

### 1.8.2. LEDGE WIDTH

A purely geometrical calculation of the maximum radius of the $i$ th ($hkl$) ledge constituting a field-ion pole assumes (Fortes and Ralph, 1967) a locally spherical end form as follows. See fig. 1.18.

$$R^2 = \rho_i^2 + (R - id_{hkl})^2$$
$$= \rho_i^2 + R^2 - 2Rd_{hkl}i + i^2 d_{hkl}^2$$
$$\therefore \qquad \rho_i \approx (2Rd_{hkl}i)^{\frac{1}{2}}. \qquad (1.37)$$

This formula is accurate when the last atom of the topmost ledge has just been removed by field evaporation.

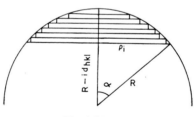

Fig. 1.18

### 1.8.3. TIP TAPER ANGLE

The tip taper angle (i.e., the semi-angle of the enveloping cone) can be estimated from field evaporation data; such a procedure might be useful if a suitable electron microscope and specimen holder (see appendix 2) were not available. The tip radius $R$ is plotted against the amount of material removed from the axis by field evaporation. Figure 1.19 shows an example of a plot made for the evaporation of 200 planes of a tungsten specimen with a taper angle of 23°.

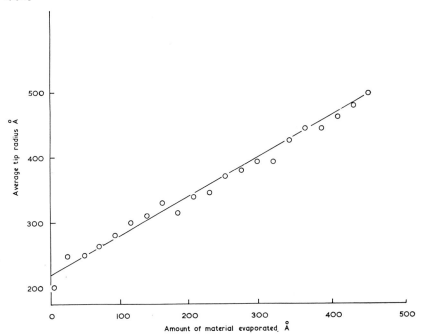

Fig. 1.19.    The variation in the specimen radius with amount of material evaporated, for a tungsten specimen imaged with helium (78 °K).

Assume that the tip is a spherical cap on a truncated cone: see fig. 1.20. Then, using the symbols defined in that figure:

In $\triangle$ OCD

$$\sin \omega = \frac{R_2}{R_2 + Nd_{hkl} + k}$$

and in $\triangle$ OBA

$$\sin \omega = \frac{R_1}{R_1 + k}$$

$$\therefore \qquad k = R_1 \frac{(1 - \sin \omega)}{\sin \omega}$$

and

$$Nd_{hkl} = \frac{(R_2 - R_1)(1 - \sin \omega)}{\sin \omega}$$

$$= \frac{\delta R (1 - \sin \omega)}{\sin \omega}. \tag{1.38}$$

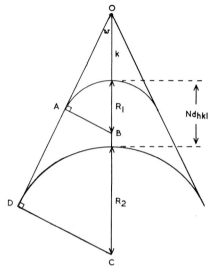

Fig. 1.20.

1.8.4. LINE OF A DEFECT

The geometrical argument developed below enables the inclination of a line defect to the axis of a field-ion specimen to be calculated approximately from field evaporation and crystallographic data. A number of simplifying assumptions are made:

(a) the radius of the tip does not change during the relevant field evaporation,
(b) equal thicknesses are field evaporated from all areas of the tip surface; this is related to (a),
(c) the field-ion image is a stereographic projection,
(d) the tip is a spherical cap,
(e) the points of intersection of a line defect with a field-ion tip, which is subject to stresses arising from the imaging conditions, are meaningful quantities.

Assumption (a) is found experimentally to be reasonable since a particular dislocation line seldom stays in the imaged area of a tip, whilst a shell more than say 200 Å thick is removed by field evaporation. (c) is true within about 10% for symmetrical specimens of most materials.

If a line defect emerges in poles inclined at angles $\varphi_1$ and $\varphi_2$ to the tip axis after field evaporation of $N$ planes of spacing $d_{hkl}$ and the tip radius is $R$, then the angle of inclination, $\alpha$, of the line defect to the specimen axis can be found as follows.

Any line in a spherical cap lies in a plane containing or parallel to the axis of rotational symmetry. It is convenient to work in this plane; fig. 1.21 is a longitudinal section through a field-ion tip to show the points of intersection of a line defect, initially and after the field evaporation of $N$ ($hkl$) planes.

Simple geometry in triangle $XYZ$ shows that:

$$\tan \alpha = \frac{\rho(\sin \varphi_1 - \sin \varphi_2)}{N d_{hkl} + \rho(\cos \varphi_1 - \cos \varphi_2)} \qquad (1.39)$$

where $\rho = R \cos \beta$, where $\beta$ is as shown on fig. 1.22.

$R$ can be calculated using eq. (1.36) following Drechsler and Wolf (1958). $\varphi_1$ and $\varphi_2$ can be measured directly on the micrographs subject to a small uncertainty (usually $<1°$) since the spiral indicating the presence of a dislocation does not necessarily begin precisely at the point of emergence of the dislocation (Smith et al., 1968). A further difficulty arises when the dislocation is dissociated; in an example in § 5.6 we have taken the apparent centre of the stacking fault as the relevant point. $\beta$ can be measured after constructing the small circle RS (fig. 1.22) by standard means.

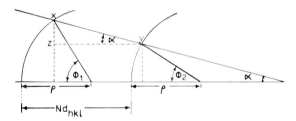

Fig. 1.21.   Longitudinal section through a field-ion tip to show the position of intersection of a line defect, initially and after the field evaporation of $N$ ($hkl$) planes.

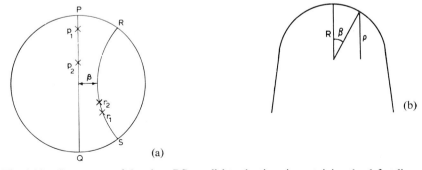

Fig. 1.22.   Stereogram of the plane RS parallel to the tip axis containing the defect line.

It is possible to obtain similar but more complex expressions for $\alpha$, which take into account changes in radius and depend on the taper angle of the specimen. However, such expressions are of academic interest, in view of the small amount of field evaporation during which a dislocation remains visible in a field-ion tip.

### 1.8.5. PLANE OF A DEFECT

A similar analysis to the above (§ 1.8.4) can be used to find the in-clinication of a plane defect to the tip axis (Hren, 1965; Morgan and Ralph, 1968). However, there is a much more elegant method.

A planar defect e.g., a grain boundary intersecting a field-ion tip surface will project as a small circle (or, as a special case, a great circle). The plane of the defect can be found simply from any single micrograph. Figure 1.23 illustrates the method: the trace of the defect is transferred to a stereo-graphic projection and the small circle is completed, if necessary by extending it outside the primitive. The pole of the small circle can then be found by standard construction (e.g. by joining two opposite ends of the diameter to the South Pole, and projecting a line which divides the subtended angle at the South Pole in half, and cuts the small circle diameter at its pole). The pole is then the normal of the defect plane, which can thus be identified.

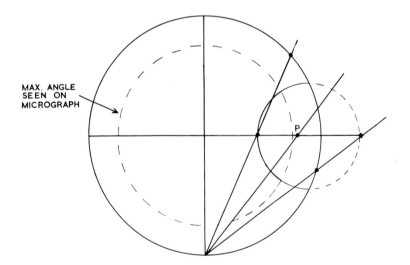

MAX. ANGLE
SEEN ON
MICROGRAPH

P

Fig. 1.23.   A method of determining for a plane defect, its indices and angle of inclination to the tip axis from a single micrograph (see text).

The angle of inclination of the defect plane normal to the tip axis is the angle from the pole to the centre of the stereogram.

As an example fig. 1.24a is a micrograph of a tungsten specimen containing a grain boundary which intersects the imaged surface. The trace of the grain boundary was plotted on the stereogram fig. 1.24b and the pole of the grain boundary plane determined by the above construction.

### 1.8.6. RELATION BETWEEN NUMBER OF IMAGE SPOTS AND RADIUS

It is useful to know the size of the sample involved when making quantitative determinations of point defect concentrations. This can be determined by the tedious method of counting but it is not necessary to make a count on *every* micrograph since there is a simple relation between the number of sites in any crystallographic region and the tip radius. Usually point defects are only investigated in specific regions of images free from

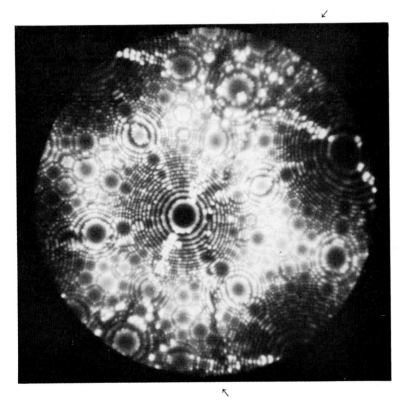

Fig. 1.24a (Courtesy B. Loberg and H. Nordén.)

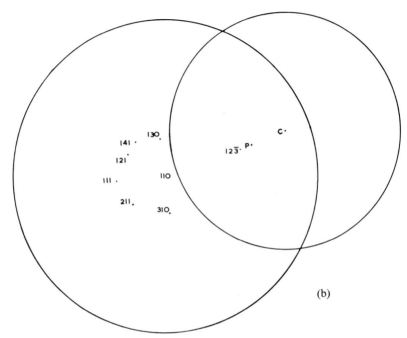

P is the pole of the small circle (boundary trace) and C its geometrical centre.

Fig. 1.24. (a) Micrograph of a tungsten specimen showing a grain boundary which makes a trace that is approximately a stereographic small circle. (b) Stereogram showing the trace, transferred from (a), of the grain boundary and the pole of the boundary plane found by the construction illustrated in fig. 1.23.

artefacts. In fig. 1.25(a), assuming the specimen tip to be a spherical cap of varying radius:

$$\sin \alpha = \frac{R_0}{R_0 + N_0} = \frac{R}{R + Nd_{hkl} + N_0}$$

where $\alpha$ is the tip taper angle, $R_0$ and $R$ are the radii of the tip before and after the evaporation of an amount $Nd_{hkl}$ from the specimen axis, $N_0$ is as shown in fig. 1.25(a).
Thus,

$$R = \frac{Nd_{hkl}}{\operatorname{cosec} \alpha - 1} + R_0. \tag{1.40}$$

We can assume that a particular crystallographic region of the tip subtends a solid angle $\Delta\theta$ at the centre of the spherical cap; see fig. 1.25(b).

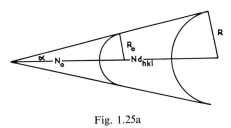

Fig. 1.25a

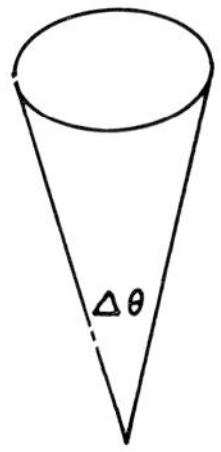

Fig. 1.25b

Fig. 1.25.

Then the surface area of a particular region is $R^2 \, \Delta\theta$ and the number of sites, $S$ is $kR^2$ where $k$ is a parameter which is a constant for one particular crystallographic region. It follows that $dS/dR = 2kR$. Substituting from eq. (1.39) gives:

$$\frac{dS}{dR} = 2k \left( \frac{Nd_{hkl}}{\mathrm{cosec}\,\alpha - 1} + R_0 \right) \tag{1.41}$$

i.e. the number of sites is a linear function of the number of planes field evaporated.

The linear relationship was also found empirically by Jeannotte and Galligan (1970), who used the linear dependence to measure the total number of sites scanned in field evaporation sequences by counting the actual number of sites on only the middle 15% of micrographs from each sequence. In this way appreciable savings can be made in the time needed to analyse micrographs.

# 2 | PRACTICAL ASPECTS OF FIELD-ION MICROSCOPY

## 2.1. Introduction

The basic design of the field-ion microscope is extremely simple and the early microscopes were not very different from the schematic drawing of fig. 1.1. The first field-ion microscope with provision for cooling the specimen was described by Müller in 1958. Vacuum joints were of the greased cone type and it operated with a background pressure of $10^{-6}$ Torr obtained by using a rotary pump plus oil diffusion pump system. There are still many microscopes operating which employ very similar designs, although not many now use greased cone joints, because of vacuum problems.

## 2.2. The vacuum system

A conventional field-ion microscope operated with helium as the image gas requires a background pressure better than $10^{-6}$ Torr in order to prevent contamination, and in some cases corrosion, of the specimen surface. Helium ionizes appreciably at a field of about 4.5 V/Å which is significantly greater than the field required to ionize all impurities. This *should* mean that at normal rates of supply no impurity atoms will be able to approach the specimen tip directly (but see § 1.4)*.

When an image gas of lower ionization potential is used, however, the electric field required for imaging is lower and it is found that impurity

---

* There is some evidence from experiments using a time-of-flight mass spectrometer, that there is contamination on the surface of specimen tips even at helium image field. This may arrive by diffusion from the specimen shank.

atoms contaminate the surface. Hydrogen for instance ionizes at a field of about 2.3 V/Å, about the same field, for instance, as that needed to ionize nitrogen which is a common background contaminant in vacuum systems. In the case of hydrogen it is found that a field induced corrosion reaction occurs with some materials (particularly iron), and that the rate of reaction depends on the pressure of contaminants. For this reason work is only possible at a vacuum of at least $5 \times 10^{-8}$ Torr (but $10^{-10}$ Torr is preferable).

Contamination is only prevented so long as the imaging field stops foreign atoms approaching the specimen surface. For experiments, such as the study of vapour deposition, that involve removing the field, it is necessary to have the best possible vacuum to prevent contamination. It is worth remembering that a monolayer of nitrogen impurity would build up at room temperature in 1 sec at a pressure of $10^{-6}$ Torr, 10 sec at $10^{-7}$ Torr, 100 sec at $10^{-8}$ Torr, and so on (the rate of build up will be somewhat slower at lower temperatures). Figure 2.1 illustrates the effects of contamination pressure on image quality.

The simplest vacuum system is an oil diffusion pump backed by an appropriate rotary pump. With the aid of liquid nitrogen contaminant traps, vacua of $10^{-7}-10^{-9}$ Torr can readily be obtained. Using mercury charged diffusion pumps it is possible to attain vacua better than $10^{-10}$ Torr though it is necessary to keep a constantly full liquid nitrogen trap above the diffusion pump in order to prevent mercury vapour from entering the vacuum system.

There is also the possibility of using vacuum systems consisting of an ion pump together with a roughing pump to evacuate field-ion microscopes. Ion pumps combined with sorbtion pumps have the advantage of combining a tremendous initial pumping speed with a very good ultimate vacuum. It is sometimes convenient to use an additional small mercury diffusion pump to remove the image gas, rather than relying on the ion pump to do this.

The image gas is generally obtained as spectroscopically pure gas sealed in glass flasks, which can be joined on to the system via a needle valve. Another common method of admitting the image gas is by means of a suitable gas diffuser. Diffusion of image gas through a heated thimble of porous material (such as quartz for helium or nickel for hydrogen) enables very pure gas to be admitted to the system without contamination (e.g. that from valves); furthermore the diffusion materials are chosen to be specific to the desired gas and result in further purification. When the required pressure is achieved the thimble is allowed to cool.

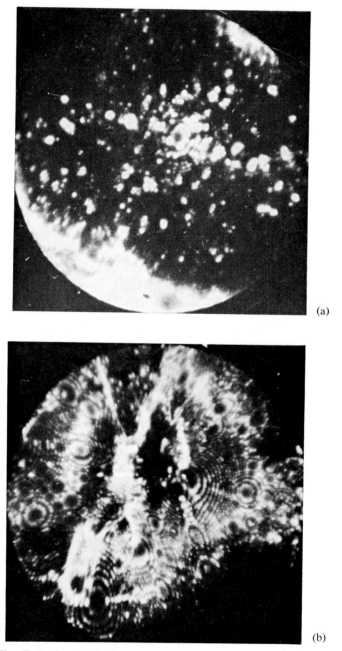

(a)

(b)

Fig. 2.1.   The effects of contamination pressure on image quality; two stages in the corrosion observed when tungsten is imaged in the presence of $10^{-4}$ Torr of oxygen. (Courtesy
G. K. L. Cranstoun.)

For the less rigorous applications, e.g., helium imaging of tungsten, it is possible to use 98% pure gas further purified by passing through a charcoal trap cooled by liquid nitrogen.

## 2.3. The microscope chamber, specimen holder and screen

The specimen must be mounted on insulated high tension leads and maintained at as low a temperature as possible. A compromise has to be reached between thermal shielding and the field of view on the screen. Experimentally it is found that a suitable arrangement is for the tip to subtend an angle of 60° at the screen which results in an image of diameter about 110°.

There is a general trend towards using stainless steel microscopes which are safer and more rugged than those constructed from glass. Problems encountered in the development of stainless steel microscopes, apart from

Fig. 2.2.　A commercially available field-ion microscope.
(Courtesy Vacuum Generators Ltd.)

their appreciably higher cost, are those of insulating the specimen holder and microscope chamber from the high tension and sometimes the outgassing of hydrogen from the stainless steel, and leaks from faulty welds. The insulation problems are often overcome by using a glass specimen holder in conjunction with a stainless steel chamber. A potentially better solution is that used by Müller (1962) and later in the 20th Century Electronics Limited Microscope, of using a stainless steel specimen holder but making the bottom from a material which conducts heat but not electricity (e.g. Lavite). In cases where the design of the microscope may later need to be modified for new applications, a glass-bodied microscope may prove more suitable since alterations such as the addition of new parts can readily be made.

It is sometimes desired to anneal the specimen *in situ*; this can be achieved by mounting the specimen on resistance wire through which a current can be passed. The temperature of the wire is monitored by measuring the resistance. A four pin specimen holder is required and the electrical connections are indicated in fig. 2.3. Müller (1949) has estimated that the temperature

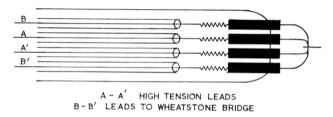

A – A′   HIGH TENSION LEADS
B – B′   LEADS TO WHEATSTONE BRIDGE

Fig. 2.3.   A four-pin specimen holder design enabling controlled specimen heating. In order to heat the specimen a current is superimposed on the high tension across AA′ and monitored with BB′. Heating with the field on is possible if an isolating transformer is used.

differences between the heater wire and the tip are 2 deg C at 1200 °K, 20 deg C at 2000 °K and 86 deg C at 2800 °K.

There is no general agreement on the best position for the fluorescent screen, which can be placed in either a horizontal or a vertical position. The microscopes that have been built up to the present time, divide themselves more or less evenly between the two possibilities. When a microscope is fitted with a horizontal screen it is necessary to examine the image with the aid of a mirror inserted under the screen. This is a major disadvantage since the light level, which may already be so low as to be on the very limit of perception, is reduced further by up to 50%.

The advantage that the vertical screen possesses in respect of direct observation needs to be weighed against the greater difficulty in recording the image. The screen is often photographed directly with a camera fitted with $f\,1$ optics, and the depth of focus of such an arrangement is so small that the camera must be accurately positioned within fine tolerances. As a result, the camera must be firmly fixed on a rigid mount with the lens to screen distance kept constant. An additional disadvantage of the vertical screen arrangement is the more complicated internal layout that is required if the specimen is to be mounted horizontally since the specimen must point towards the screen. Despite the disadvantages it is, however, sometimes preferable to use the vertical screen, for instance if an external image intensifier is to be used with the microscope.

## 2.4. The image gas

Probably the simplest image gas to use is helium which ionizes at fields above about 4.5 V/Å, and therefore can only be used to image those materials that require a field greater than 4.5 V/Å for evaporation. At liquid neon temperature these include iridium, molybdenum, niobium, platinum, rhenium, rhodium, tantalum and tungsten and it is also *just* possible to obtain a stable image from iron, nickel, palladium, titanium, vanadium and zirconium (Müller, 1965). Most alloys of the above elements and some ceramics will also be stable.

The range of materials that may be studied with the microscope can be extended by using an image gas with a lower ionization potential. The values of the image field for a number of possible gases are listed in table 2.1.

TABLE 2.1

Data on field-ion microscopy with various gases (Müller, 1965)

| Gas | Ionization potential (eV) | Image field (V/Å) | Potential resolution for 500 Å radius tip (Å) |
|---|---|---|---|
| He | 24.6 | 4.50 | 1.2(20°K) |
| Ne | 21.6 | 3.70 | 1.3(20°K) |
| $H_2$ | 15.6 | 2.28 | 1.6(20°K) |
| A | 15.7 | 2.30 | 3.2(80°K) |
| Kr | 14.0 | 1.94 | 3.5(80°K) |
| Xe | 12.1 | 1.56 | 4.0(80°K) |
| $N_2$ | 15.5 | 2.26 | 3.3(80°K) |
| $O_2$ | 12.5 | 1.64 | 3.8(80°K) |
| Hg | 10.4 | 1.24 | 8.5(300°K) |
| Cs | 3.9 | 0.28 | 21.0(400°K) |

The only image gases, other than helium, which are commonly used are neon and hydrogen. Argon has also been used on occasions although the images can suffer from low resolution but Arthur (1964) has obtained some micrographs of germanium and Turner (private communication) has ascertained the potential resolution, image field etc. for aluminium with argon image gas, and obtained argon-ion images of tungsten of comparable quality to those obtained with hydrogen-ion imaging. Neon suffers from two disadvantages as an image gas. There is a very considerable reduction in intensity using neon for direct imaging both because the phosphor efficiency is much less than for helium ions, and because of the reduced operating pressure that is required due to the greater cross-section for gas scattering. In addition the heavy neon ions bombarding the specimen surface during thermal accommodation may induce lattice defects (Müller, 1964), which could rule out the use of neon for the study of radiation damage and in particular for determination of vacancy concentrations. The first of these two disadvantages disappears if the microscope is fitted with an image converter (see § 2.6) to convert the ion image into an electron image, since the heavier neon ions will produce more secondary electrons than, say, helium ions.

Hydrogen was the image gas used by Müller in his original experiments, and owing to its very low ionization potential might be expected to be amongst the most versatile of all the gases listed in table 2.1. However, the low ionization potential also means that it is difficult to keep the surface free from contamination. The common impurities present in high vacuum systems (CO, water vapour, etc.) are able to reach the specimen tip surface at the fields necessary for hydrogen ionization. The constant arrival of impurities causes undesirable surface reactions, and corrosion (uncontrollable field etching*) may result.

The use of hydrogen, therefore, was virtually neglected for almost ten years until the introduction of ultra-high vacuum microscopes.

Müller et al. (1965) have demonstrated that excellent results can be obtained using mixtures of image gases (see fig. 2.4). In particular, the results obtained on iron whiskers were very good using helium with a trace of hydrogen as the image gas. The exact role that the hydrogen plays is not entirely clear, though Müller et al. suggest that an invisible layer of hydrogen is adsorbed on the surface which improves the accommodation of helium

---

* It is possible to distinguish field evaporation from field etching by noting the greater field dependence of the rate in the case of field evaporation (Brandon, 1966a).

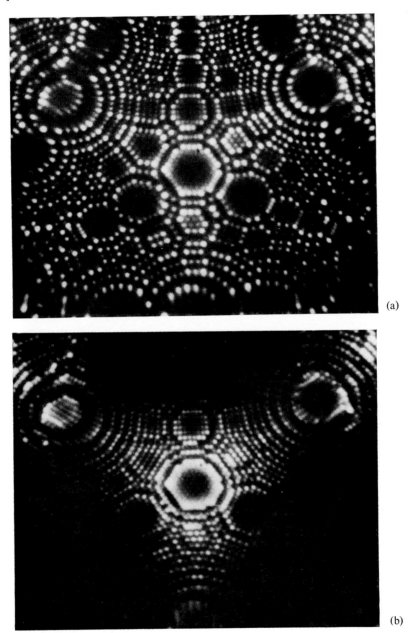

(a)

(b)

Fig. 2.4.   A hydrogen-promoted image of tungsten: (a) helium imaging, no promotion, (b) helium imaging with hydrogen promotion. The resolution *within* the (111) plane is improved. (Courtesy E. W. Müller.)

and reduces the image field, and in fact Turner (private communication) has detected the presence of an adsorbed surface layer during hydrogen ion microscopy. Brandon (private communication) feels that the effect of the hydrogen is to preferentially evaporate interstitial impurities, allowing a perfect surface to develop. Müller has suggested, that the discovery could mean a breakthrough for field-ion microscopy of the common transition metals, which previously could not be imaged very well with helium.

## 2.5. The specimen coolant

It is desirable to operate the microscope with the specimen at as low a temperature as possible. The theoretical lower limit must be the temperature at which the image gas will become adsorbed on the specimen surface, but in practice the lower limit is more often set by considerations of cost and convenience. The resolution of the image will evidently be dependent on the temperature because the lateral velocity component of the ions is the major limiting factor in determining the resolution. The potential resolution is approximately (Nishikawa and Müller, 1964):

$$\delta = \frac{6 \times 10^{-4} Tr}{F} \text{ Å} \tag{2.1}$$

where $T$ is the tip temperature in degrees absolute; $r$ is the specimen radius in Ångstrom units; and $F$ is the field strength in volts per Ångstrom. In practice the resolution is rather poorer but the temperature dependence is maintained, or even slightly exaggerated.

The use of lower specimen temperatures also increases the range of materials which may be studied, since the field at which evaporation commences for any particular material remains nearly constant as the temperature varies, but the minimum field for ionization is lowered by a reduction in temperature (Southon and Brandon, 1963). The result is that the "working range" (§ 2.6) of the material is increased and in some cases it is possible to obtain stable images from "difficult" materials simply by lowering the temperature.

Possible specimen coolants include solid carbon dioxide (194.5 °K), solid iso-pentane (113 °K), liquid nitrogen (77.4 °K), solid nitrogen (63.2 °K), liquid neon (27.1 °K), liquid hydrogen (20.4 °K) and liquid helium (4.2 °K). Liquid nitrogen is the most commonly used coolant because of its wide availability, safety and relative cheapness. Solid nitrogen is produced *in situ* by evaporation of liquid nitrogen under reduced pressure. The stability obtained using solid nitrogen as specimen coolant with neon as the imaging gas can be comparable with that achieved using liquid hydrogen and helium gas. There is a considerable fire and explosion risk involved in using liquid hydrogen, the coolant used by Müller (1960). Liquid helium has an extremely low latent heat, and there are very special problems involved in constructing a microscope to operate using it as a coolant (Attardo and Galligan, 1967; Forbes and Southon, 1966). To gain full advantage of the use of liquid helium, as with any other coolant, it is essential that care is taken to ensure good thermal contact between the specimen and coolant; where this is done the resolution is outstanding: fig. 2.5.

Liquid neon has been used by Brenner (1962) and by Bowkett and Ralph (1966). It has a very high initial cost* but there is a generous rebate

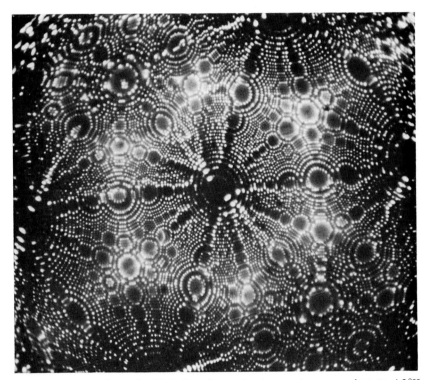

Fig. 2.5. Tungsten imaged with helium ions, at a temperature very close to 4.2°K. (Courtesy R. G. Forbes.)

* In 1970 $100–$150 per litre.

on recovered gas which can amount to 65% of the initial cost and since liquid neon has a high latent heat it can therefore be an economic proposition.

An alternative to the use of bulk liquid coolants is to use a miniature Joule-Thompson Liquifier (e.g. the "cryo-tip" marketed by Air Products or a similar device made by Hymatic). These devices fit in a field-ion specimen holder (see fig. 2.6) and using gas under pressure, make a small quantity of liquid hydrogen or helium which can be handled safely in a closed system.

Fig. 2.6.   Joule-Thompson miniature gas liquifier.

The cooling capacity is up to about 6 Watt. They have an additional advantage because they enable the coolant temperature to be varied over the range $4°K–78°K$ quite easily, and at some temperatures they may also have an economic advantage despite a high rate of gas consumption.

One other method of specimen cooling is to use a continuous flow of cold helium gas; specimen temperatures from room temperature down to nearly $4°K$ can be achieved (e.g. Klipping and Vanselow, 1967; Seidman et al., 1969). The cold helium gas is obtained by boiling-off liquid helium, and the temperature is controlled by varying the rate of boil-off and hence the gas flow. Pimbley and Ball (1965) have demonstrated the use of a Collins helium refrigerator to cool the specimen. With this system it is possible to control the specimen temperature between 10° and 35 °K.

## 2.6. Image intensity

The number of ions projected from each ionization centre is about $10^3$–$10^4$ per second. Since heavy particles such as ions have a low phosphor efficiency, the inevitable result of the low ion-current is a faint image. This constitutes one of the chief disadvantages of the microscope and means that without some form of image-intensification, photographing the image can be a lengthy procedure. With $f$ 0.95 optics and the fastest films available that have an acceptable grain-size, exposure time ranges between 1 hr with the specimen at 3–4 kV to 10 sec at 30 kV, using helium as the image gas.

Neon, which is also used as an image gas, has a lower phosphor efficiency, and exposures can be ten times as long. The image intensity also depends on the image gas pressure, which is limited by image-blurring due to gas scattering; because of its larger cross-section blurring occurs at much lower pressures for neon than for helium, which aggravates the problem. The pressure of the image gas is also limited in practice (with some microscope designs) by the need to minimise the boil-off of refrigerant.

The applied voltage at which the image is examined and photographs are taken is known as the "best image voltage" (BIV) and depends on the tip radius. In reality, each plane in the image has its own BIV due to the varying work function, local geometry, and possibly other factors, and an overall compromise must be chosen. At too low a field whole planes may not show up, whilst at too high a field, blurring of the image takes place. Since the image intensity depends on the applied voltage, the highest useful field is normally chosen.

The image intensity can be thought of in terms of the current of ions arriving at the screen, and accurate ion-current measurements have been made by Southon and Brandon (1963). Using a helium field-ion microscope, current-voltage characteristics were measured as functions of tip radius, operating temperature and helium pressure and also for specimens of various metals. Figure 2.7a shows the current-field characteristics for various metals at 77 °K; the "working range" of a specimen can be defined on this diagram as the region BD where field-evaporation occurs at D, and the point B is known as the "cut-off" and corresponds to the minimum image brightness useful for unamplified viewing or photography.

There is an advantage in having a small working range; the progress of field evaporation can be followed more readily. Where the working range is too large, however, some or all of the image may become blurred as the evaporation field is approached.

Practical considerations suggest an extension of the definition for "working range". As mentioned in § 1.6 materials may yield under the stress as the imaging field is applied, and this will further restrict the working

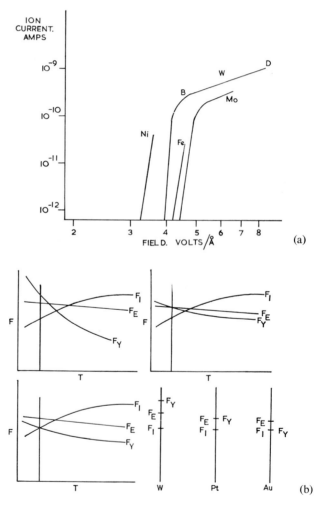

Fig. 2.7. (a) The current-field characteristics for various metals at 78 °K. (Courtesy M. J. Southon). (b) Schematic diagrams to show the temperature dependence of $F_i$, $F_e$ and $F_y$ and their relative values at 78 °K for three metals.

range; see fig. 2.7b. In future we shall use the term *Working Range* to describe the useful region where stable images are obtained.

Various attempts have been made to reduce the photographic exposure times by using some form of image intensification, Brandon et al. (1964) attempted to convert the ions directly into electrons inside the microscope, by using a fine mesh, aluminium-coated grid inserted between the specimen and the screen as an image-converter. The method, first suggested by Von Ardenne (1956), is to form the ion image on the fine metal mesh and collect the secondary electrons through the holes by applying a large positive field to the far side of the mesh. An electron image will be more intense because the phosphor response (i.e. the number of photons released per impact) is greater for electrons than for helium ions up to about 12 keV, while the response for neon and argon ions is well below that for electrons at all working voltages (see fig. 2.8). The secondary electron image obtained by Brandon et al. (1964) was more intense than a helium ion image even when the image gas was neon or argon. However, there were a number of problems including loss of resolution, image doubling and image "flares" due to secondary emission processes. Brandon (1966c) has outlined the experimental

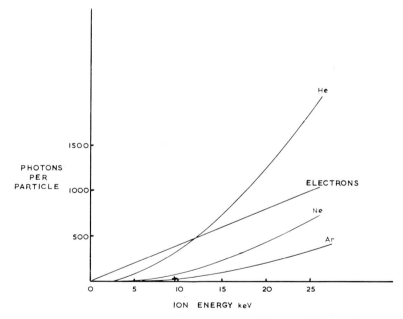

Fig. 2.8.   Phosphor response for various particles as a function of voltage between the specimen and the screen. (Brandon et al., 1964.)

procedures involved, and described a number of improvements including the introduction of a biasing field to limit the effective energy spread of the electrons.

Turner and Southon (1967) improved on the simple "proximity focussed" converter system, using magnetic focussing to reduce the loss in resolution due to the transverse momentum of the secondary electrons. Their system, illustrated in fig. 2.9 has performed well, giving a resolution in excess of 50 line pairs/mm, and a reduction in photographic exposure time of about $10^4$ for a neon-ion image.

It is recommended that for applications where it is necessary to use

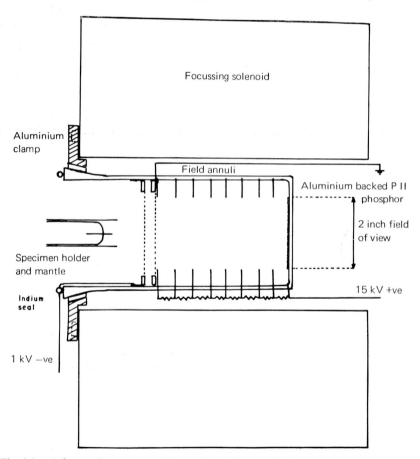

Fig. 2.9.   A focussed converter grid intensifier. (Courtesy P. J. Turner). The meshes are shown dashed. A similar system can be used for a channel-plate converter by replacing the meshes with a channel-plate.

neon or argon as the imaging gas, an image converter system should be used.

A promising alternative to the various mesh converter systems is the channel-plate converter (Turner et al., 1969) which consists of a bundle of coated fine hollow fibres across which a potential is applied. An ion striking a fibre at one end generates electrons which are accelerated down the tube and multiplied by further collision with the walls. The emerging amplified beams of electrons are focussed magnetically onto a phosphor screen. The photographic exposure time is reduced by a similar factor to that for mesh converter systems. Because of their military applications, in some countries unfortunately the most developed versions are not readily available. Channel-plates are available from Mullard (Redhill, Surrey, England) and from Bendix (Cleveland, U.S.A.). Also, Lewis and Gomer (1969) have reported on the performance in argon-ion microscopy of a channel-plate, 5 cm in diameter, obtained from the Rauland division of the Zenith Corporation (U.S.A.).

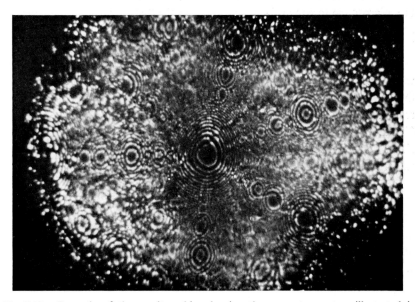

Fig. 2.10. Example of the results achieved using the converter system illustrated in fig. 2.9: a specimen of nickel. A grain boundary and a number of dislocations are visible. (63 °K, neon image gas). (Courtesy P. J. Turner.)

Brandon et al. (1964) also investigated the possibilities of post acceleration of ions. Figure 2.11 shows the experimental arrangement that they used both for the post acceleration and the image conversion experiments; for post acceleration the first grid was omitted and the second grid was maintained at earth potential with the screen charged negative. They used 70% transmission grids with 200–350 squares per linear inch (obtained from E.M.I., and found a significant improvement in the light output from the fluorescent screen (e.g., a 2 min photographic exposure reduced to 15 sec).

Other attempts at internal image intensification have used devices that increase the ion current at the screen. These devices have the advantage of being simple to construct, but there is a limit on their usefulness because the rate of ion damage of the screen phosphor increases with the ion current (but is largely independent of ion energy). An example of such a device is the "dynamic gas supply" used by Waclawski and Müller (1961) and by Ryan and Suiter (1965) where the gas pressure at the tip was increased by completely enclosing the tip in the cathode assembly and reducing the dimensions of this assembly.

The most successful external image-intensification system used between the microscope screen and camera has been the 3-stage "cascade" intensifier system, and Müller's results with an R.C.A. system have been particularly impressive (e.g. Müller, 1962; McLane et al., 1963). Whitmell and Southon (1965) attempted to use a 5-stage dynode intensifier but the effective gain which can be obtained with this type of tube was found to be severely limited by signal-induced optical "noise", resulting in loss of contrast and resolution. Experiments with cascade-type intensifiers made by E.M.I. Ltd., (4-stage) have been encouraging. Dynamic events in the intensified image can either be recorded using a ciné-camera (Müller, 1962) or using a television camera and video-tape recorder (Cranstoun, 1968).

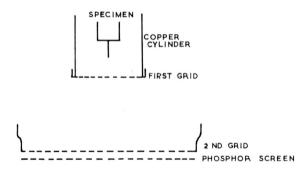

Fig. 2.11.   The experimental arrangement used by Brandon et al. (1964) for post acceleration and image conversion.

## 2.7. Experimental techniques

### 2.7.1. SPECIMEN PREPARATION *(see also appendix 3)*

Specimens are best prepared from material in the form of a wire or at least an axially symmetrical blank. Hence the first stage of specimen preparation is to obtain the material in wire form if possible, by drawing or grinding. In cases where this is not possible, e.g. titanium carbide, tungsten-silicon alloys; it is usual to cleave or to spark machine something approximating to a wire from a single crystal or a solidified drop of material.

The actual tip for field-ion microscopy is generally prepared from the wire by electropolishing at an interface normal to the wire axis. Sometimes this interface is the electrolyte/air interface in which case the wire and the counter electrode are simply dipped into the electrolyte. More commonly, however, with this system the attack at the electrolyte/air interface is not sufficient to ensure the highly localised attack that is desirable for the preparation of a field-ion tip. In such cases the most common system used is where a thin (3–4 mm) layer of electrolyte is floated on an inert liquid and the polishing conditions adjusted to give rapid dissolution at or near the liquid/ liquid interface. Suitable dense liquids are carbon tetrachloride or 1, 1:2, 2 tetrabromoethane which is denser and is suitable for use with phosphoric acid based electrolytes.

AC or DC electropolishing is used, as found experimentally more suitable (see appendix 3). Polishing is usually fairly rapid taking a matter of minutes, and it is advisable for the course of the electropolishing to be followed by constant observation of the wire through a telescope, so that

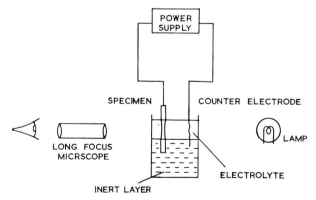

Fig. 2.12. The arrangement for electropolishing using the "thin-layer" technique. The power supply between the specimen and the counter electrode may be AC or DC as required.

polishing may be stopped once the piece of wire below the interface has dropped off. In cases where the electrolyte is opaque and it is not possible to follow the electropolishing in this way, a simple circuit of the type described by Morgan (1967b) can be constructed to stop the electropolishing when the current falls rapidly as the bottom piece of wire drops off. Figure 2.13 shows stages of specimen preparation.

It is conventional to polish back a very small amount, although this tends to blunt the tip, in order to remove the highly deformed region where the wire fractured, and to avoid too slender a specimen. A good specimen is usually one that under a low magnification microscope looks like a well sharpened pensil with a uniform taper of about 15°. The severe deformation at the specimen tip is due to the weight of the bottom piece of wire hanging,

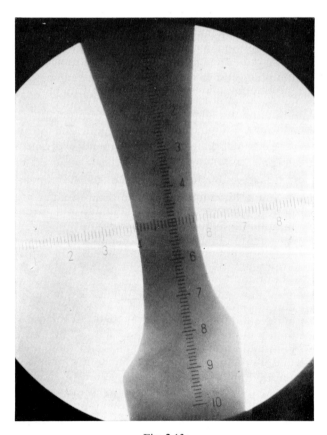

Fig. 2.13a

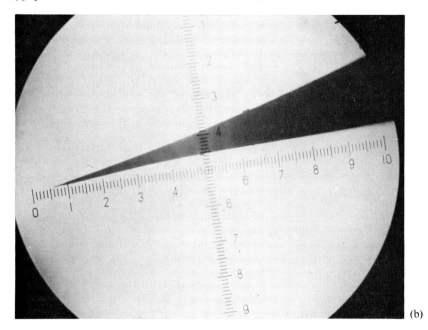

(b)

Fig. 2.13.   Stages in the electropolishing of a specimen: (a) formation of a waist (b) the
final polished tip (each small division of the graticule is 10 $\mu$m).

at the end of the electropolishing, on a very small cross-section, which will
result in stresses of $10^{10}$–$10^{11}$ dyne cm$^{-2}$, i.e. almost the theoretical strength
of the structure, and as a result dislocations may be present in large numbers.
In addition this region *may* have been heated and quenched during electropolishing. For these reasons backpolishing is highly desirable.

Müller (private communication) has devised a variant of the usual
method of specimen electropolishing which does not involve fracture. The
specimen wire runs part-way into a capillary tube immersed in electrolyte;
the capillary apparently modifies the throwing power of the system so that
a point is produced directly at the end of the wire.

After electropolishing, it may be necessary to clean the specimen in
order to remove deposits left from electropolishing. These deposits are not
always immediately obvious in the field-ion microscope, although they may
give rise to various curious and apparently inexplicable effects. However,
they are immediately obvious if the tips are examined in an electron microscope, where the deposits may result in a complete lack of penetration.
Bowkett, Loberg and Norden have found that these films can often be elimi-

nated by electropolishing in an electrolyte consisting of certain commercial photographic developers which contain a small amount of hydroquinone as an 'activator' (Fasth et al., 1967). Traces of the polishing solution should be removed e.g. by washing with distilled water followed by drying with ethyl alcohol. Some specimens may need to be stored in a dessicator to reduce atmospheric corrosion.

The conditions for specimen preparation are found empirically and are not always simple. There have been few attempts to relate the conditions to standard electropolishing or electroetching; it would seem desirable to avoid electroetching, especially towards the end of specimen preparation in order to avoid attack at grain boundaries. However, in a few cases it has been found that better results are obtained in the electroetching range; it is very common for the preparation to be done in the gas evolution range which occurs beyond polishing on a conventional anodic current/voltage curve.

One method that has been used with some success by Southworth (1968) is to polish at maximum current, which means gradually reducing the voltage as the specimen thins.

In the case of AC electropolishing, gas evolution always occurs and conditions will pass through the full range. Data on the electropolishing of thin foils for electron microscopy (e.g. Hirsch et al., 1965) can usually be adapted directly to give a method – though not necessarily the best – of preparing specimens of a 'new' material for which no data on field-ion specimen preparation are available.

Müller (1960) has successfully used various molten salts as electrolytes, but it seems that now other media are available, these will prove to be too inconvenient for general use.

Simple chemical polishing can sometimes be used; for instance, silicon and germanium can be polished in a solution known as CP4 which is an acetic acid/nitric acid/hydrofluoric acid mixture. This method would be suitable for materials like MgO or NaCl. A common method of preparing thin foils for the electron microscope is to use a chemical polish for the initial thinning, followed by a suitable electrochemical polish. This double technique could be suitable for polishing field-ion specimens, especially in cases where the electropolishing tends to be uneven e.g. a two-phase material.

Materials that can be obtained in the form of whiskers or fine fibres can be prepared by flame polishing, e.g. a silicon carbide whisker can be burned in an oxygen-propane flame and any droplets of silica removed with hydrofluoric acid to give a reasonable field-ion tip (Smith, 1969a).

There are advantages in examining specimens in an electron microscope before and after examination in a field-ion microscope. This enables specimens that have not been prepared properly to be eliminated, and also makes subsequent image interpretation very much easier. The Philips EM300 microscope has a rod type specimen holder and is easily adapted to take field-ion specimens and to allow them to be rotated axially whilst under observation. Most other electron microscopes have specimen holders which drop down into the objective lens, so that it is difficult to insert specimens with a dimension greater than 2–3 mm. Details of the construction of suitable specimen holders are given in appendix 2.

### 2.7.2. Screen preparation

The screen usually used for field-ion microscopy consists of a phosphor laid on an optically flat glass disc with a conducting surface film of tin oxide. It is possible to replace the glass disc with a fibre optic disc in order to reduce the loss of intensity and resolution involved in recording the image. Newman, Lefevre and Hren (1966) have used a $4\frac{1}{4}$ inch diameter fibre optic screen with 15 micron fibres which they obtained from Mosaic Corporation Inc. (USA) and obtained outstandingly good results by using slow, fine-grained photographic plates placed directly on the outer surface of the fibre optic screen (fig. 2.14).

The screen phosphor can be laid either by settling from solution in water or methanol, or by 'puffing' the dry phosphor powder on to the screen. The phosphor can be made to adhere to the screen evenly and more strongly either by dissolving the phosphor in nitrocellulose/amyl acetate solution (which acts as a form of glue) or by pretreating the tin oxide layer on the screen with a little 10% phosphoric acid solution in acetone, and heating the screen still in contact with the solution to 400 °C for 1 hour (Gomer, 1961). There are a number of suitable phosphors available, and it is generally found that those that give a high initial light level (e.g. activated zinc sulphate phosphor FF 12, obtainable from Levy-West Ltd) tend to decay rather rapidly as they suffer impact damage from the imaging ions. A good all-round phosphor is Willemite (Levy-West H 918), which is resistant to screen damage and is rejuvenated by low temperature annealing during vacuum bake-out. It is not possible to 'aluminize' field-ion screens by evaporating a thin layer of aluminium on to the phosphor, as is done with television screens, because an aluminium layer, although largely transparent to electrons, provides an effective barrier for the low energy ions used for field-ion imaging.

### 2.7.3. Photographic techniques

Unless some form of image intensification is used, the field-ion image

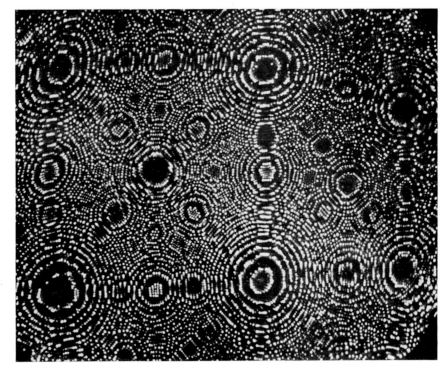

Fig. 2.14.   Ordered nickel-molybdenum, $Ni_4Mo$, photographed using a fibre optic screen.
(Courtesy J. J. Hren.)

is extremely faint and care must be taken to minimise losses in the recording system, which is conventionally a camera placed in front of the screen. The camera lens aperture preferably should be bigger than $f$ 1 and a suitable lens is made by Wray Ltd designed to be used at a demagnification ratio of 4:1. It is also possible to use a Canon 7 camera fitted with the standard $f$ 0.95 lens and a suitable extension ring for close-up work; however, this lens is corrected for infinity focussing and it would be necessary to check its resolution when used as a close-up lens. Brenner (1966) has demonstrated a simple means of increasing the effective image brightness by reducing the specimen to screen distance to about 1.5 cm which gives an image about 2.5 cm in diameter. This can be recorded on 35 mm film using two front-to-front coupled 4:1 $f$ 1 lenses giving 1:1 magnification overall. The photographic exposure time using this method is about one twentieth that for a more conventional system, but phosphor grain size tends to be a problem.

Recording films may be chosen either for speed or for resolution (fine grain size). Kodak Tri-X is commonly used where the resolution is thought to be important. Absolute speed is best obtained with various films developed for astronomical purposes, such as 103-a G (Kodak) which has been used by Müller and others. There are two Polaroid films that have been used; the fastest (10000 ASA) tends to be too contrasty but good results have been achieved using 3000 ASA Polaroid, particularly if the film has been pre-exposed to a faint light source in order to increase speed and reduce contrast.

Typical photographic exposure times, without any form of image intensification, using Tri-X film and H918 phosphor with helium as the image gas are:

imaging voltage,   6 kV; gas pressure, $6 \times 10^{-4}$: exposure 25 min;
imaging voltage, 10 kV; gas pressure, $6 \times 10^{-4}$: exposure   4 min;
imaging voltage, 10 kV; gas pressure, $2 \times 10^{-4}$: exposure 18 min;
imaging voltage, 15 kV; gas pressure, $4 \times 10^{-4}$: exposure   1 min.

It is not possible to reduce the exposure time by having the photographic material inside the microscope, as is done with electron microscopes, due to the poor vacuum qualities of photographic materials (Mulson and Müller, 1963). However, McNeill (1968) has investigated the possibility of direct electrographic recording of field-ion images (effectively a "Xerox" process). The problem is to find a suitable receptor plate material which must be able to trap the surface charges produced by the incident ions. It must also retain the positive charge pattern so produced for a sufficient time to enable the plate to be removed from the microscope and developed in a fine grained liquid electrophotographic developer of negative particle charge.

### 2.7.4. THE ATOM PROBE

Field-ion microscopes have been used in conjunction with mass-spectrometers e.g. for investigating chemical reactions. However, Müller et al. (1968) were the first to combine a field-ion microscope with a time-of-flight mass spectrometer. An experimental arrangement is shown in fig. 2.15. An otherwise conventional field-ion microscope has the tip mounted on a universal joint and a small hole in the screen opening into the mass-spectrometer flight tube. When a chosen atom is evaporated in such a way that it passes through the hole in the screen, it is possible to measure the time between the voltage pulse required for evaporation and the ion's arrival at the detector. The time-of-flight $t$ depends on the mass $m$ and charge $ne$ of the ion, since:

$$Ene = \tfrac{1}{2}mv^2$$

where $E$ is the potential between the tip and screen (the screen and detector are both earthed), and $v$ is the ion velocity taken to be constant. If $L$ is the tip to detector distance, then:

$$t = \frac{L}{v} = \frac{Lm^{\frac{1}{2}}}{(2\,Ene)^{\frac{1}{2}}}.$$ 
          (2.2)

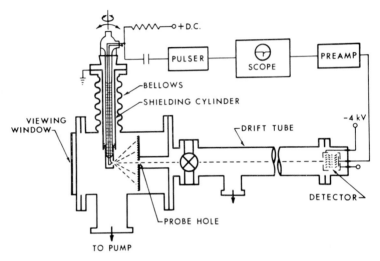

Fig. 2.15.  The experimental arrangement for an 'Atom Probe'. A field-ion microscope is combined with a time-of-flight mass spectrometer. (Courtesy E. W. Müller.)

The atom probe provides a powerful technique for the identification of individual atoms. It has obvious applications in alloy studies and in checking theories of field evaporation, since it is possible to determine the charge of the evaporating species. The main experimental difficulties are the accurate measurement of the time-of-flight on which depends the resolution, and obtaining a square voltage pulse of short duration ('square' refers to the shape of an oscilloscope trace of voltage against time and implies a short rise-time and decay-time). A further experimental difficulty lies in knowing precisely which atom goes through the probe hole, since the image gas ion and field evaporated ion may not follow quite the same trajectory (Müller et al., 1968).

# 3 | SOME COMMON ARTEFACTS

## 3.1. Introduction

Before going on to describe the interpretation of experimental results obtained with the field-ion microscope, it is necessary that we consider the contrast of some images which might tend to mislead the unwary. The microscope uses its materials in a somewhat unusual form of specimen and we need to consider carefully what defects may be introduced or eliminated as a result of the preparation technique, the high surface to volume ratio or the stress on the specimen due to the imaging field.

There are a number of possible image 'defects' which may arise from a series of trivial causes, best summed up as 'optical illusions'. Firstly dark spots on the phosphor (as a result of phosphor damage or of faulty screen preparation technique) which could be confused with vacancies unless care is taken (see § 3.2.5). A second type of optical illusion can arise as a result of reciprocity failure in the recording film: certain faint atom spots do not show up at all in the photographed image and this gives rise to a threshold effect which may also be misleading in images from alloys where there is a variation in spot intensity; see ch. 9. Another common illusion is the presence of small bright regions in a photographed image that have arisen as a result of a small discharge inside the microscope; an example is shown in fig. 3.1.

In the rest of this chapter we shall consider a number of image features, that can be subsumed under the heading of artefacts. This means that an experienced microscopist would tend to accept these features as either 'understood' or 'normal' in an image, and often capable of being almost

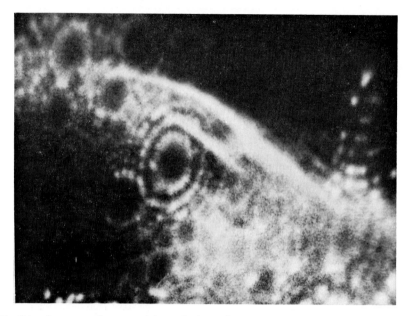

Fig. 3.1.   Image artefact caused by a discharge in the microscope during photographic recording of the image.

ignored. This does not mean, of course, that they may not be important features in themselves under the right circumstances, but just that they tend to occur in many images and often do not affect the interpretation of the feature being studied.

### 3.2. Specific examples of artefacts

#### 3.2.1. MASSIVE STREAKING OF IMAGES

It occasionally happens that a field-ion image instead of taking its more normal form has the appearance of fig. 3.2 and consists largely or solely of a series of bright streaks running across the image. This type of *massive* streaking is invariably associated with considerable image asymmetry; the specimen will have an approximately elliptical cross-section with the two principal radii of the ellipse differing by a factor of perhaps ten which can be demonstrated by examining the specimen in an electron microscope. Since the magnification is inversely proportional to the radius all the image spots become elongated parallel to the minor axis of the ellipse and any small surface ridges that happen to be present after electropolishing will give rise

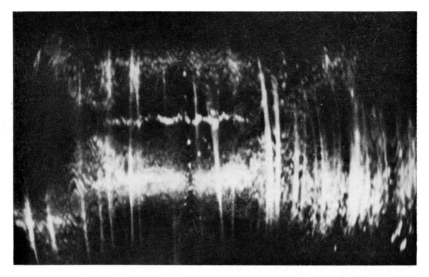

Fig. 3.2.   Image streaking caused by extreme asymmetry of the specimen. The usual iridium pattern has been distorted beyond recognition.

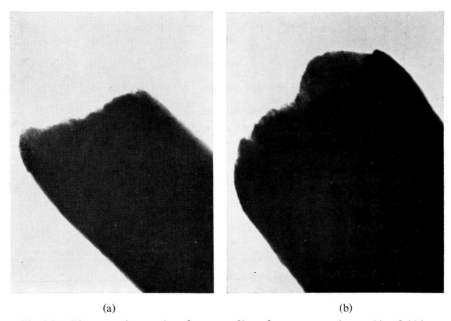

(a)                                              (b)

Fig. 3.3.   Electron micrographs of two profiles of an asymmetric graphite field-ion specimen separated by a 90° rotation. ($\times$ 50 000.)

to bright streaks. It is sometimes possible to remove this type of streak with a large amount of field evaporation. Brandon (1966d) has pointed out that an asymmetric end form such as that shown in fig. 3.4 could be stable to field evaporation providing that the principal radii were inverse functions of each other. He suggested that surface steps giving image streaking, might be stable on such a surface.

Specimen asymmetry usually arises during electropolishing and is most common when the thin layer technique (see § 2.7.1) is used. It probably arises as a result of the deformation induced by the gas bubbles that are evolved during electropolishing, combined with the high stress level that exists just before the bottom piece of wire drops off. The exact mechanism is not simple, however, for it is found that the state of the material exerts an influence. For the case of electropolishing of tungsten in aqueous potassium cyanide solution it was found by Ranganathan et al. (1965) that approximately 10% of annealed specimens showed considerable image asymmetry together with a high density of streaks, while *100%* of specimens prepared from material that had been cold-worked and then neutron irradiated showed these effects. Furthermore, the streaks have been found to be roughly parallel to recognisable crystal planes; in tungsten these are {112} and {110} with {112} being much the more common, and in fcc material (e.g. iridium) the streaks most commonly lie parallel to {111}. Knowing that in tungsten

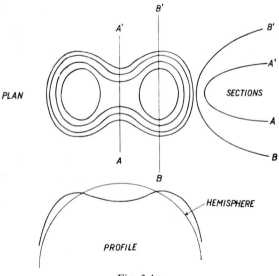

Fig. 3.4a

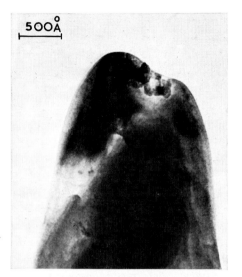

(b)

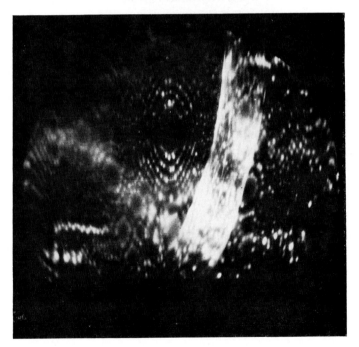

(c)

Fig. 3.4.   (a) Contour map of a field-ion specimen surface which might be expected to yield a streaked image. (Courtesy D. G. Brandon.) (b) Electron micrograph (Courtesy H. Nordén) of such a tungsten specimen, which gave the field-ion image of fig. 3.4c. (Courtesy B. Loberg.)

specimens prepared from wire, the tip axis is almost always close to $\langle 110 \rangle$ as a result of the preferred orientation developed during fabrication into wire, the major and minor axes of the specimen cross-section can be seen to be [112] and [11$\bar{1}$] respectively for (112) streaks or [110] and [001] respectively for (110) streaks (see also fig. 3.5 and 3.6).

There is apparently some connection between the amount of deformation in the polishing process, the initial state of the material and the occurrence of an asymmetric and streaked image. It is possible for instance to take some deformed and irradiated tungsten wire which always gives image streaks when electropolished using a thin layer of aqueous potassium cyanide solution and prepare symmetric specimens with an extremely low incidence of image streaking by DC electropolishing in sodium hydroxide.

Streaks very often occur, no matter what the preparation technique, if specimens are polished from wire or other starting material which is itself highly asymmetric. Ralph (1964) has found that streaks tend to occur in specimens prepared from vice deformed tungsten-rhenium wire, and Bowkett (1966) has confirmed that this also happens in vice deformed pure tungsten wire, if the vice deformation results in the wire having an oval cross-section. Similarly Smith and Bowkett (1968a) correlate the streaking in graphite field-ion images with the extreme asymmetry of the end form as demonstrated by electron microscopy; fig. 3.3.

Many otherwise perfect specimens show short sharp streaks at the edge of the image, which apparently arises from protrusions or contamination left on the specimen shank during electropolishing.

Further information on image streaking can be found in papers by Ranganathan et al. (1965), Brandon (1966d), Fasth et al. (1967), Smith et al. (1968) and Fortes and Ralph (1968a). It is interesting to note that this general image streaking is also encountered in field electron emission patterns e.g. those of germanium published by Ernst (1966). Rose (1956) gives an account of the field distortion above an asymmetric field emitter.

A further case of streak contrast occurs in images of multiple tips such as those resulting from fracture in the microscope where spot images overlap owing to image superposition. Such effects can occur around surface craters due to heavy particle irradiation damage perhaps in some cases with an additional contribution from local radius changes.

### 3.2.2. LOCAL STREAKING OF IMAGES

Defects such as stacking faults, microtwins and grain boundaries have planar interfaces across which there is a displacement which in general is

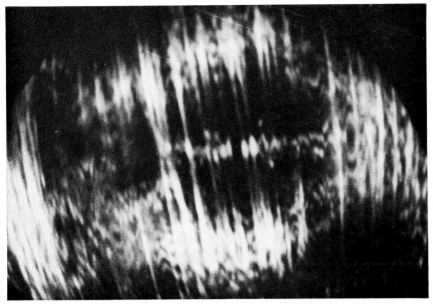

(a)

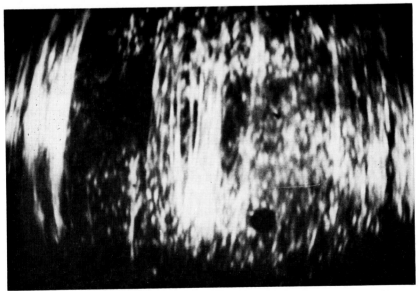

(b)

Fig. 3.5.   (a) Type I streaks on {110} planes in a field-ion image from an elliptical tungsten specimen; 78 °K, helium image gas. (b) Type II streaks on {112} planes in a field-ion image of tungsten; 78 °K, helium image gas.

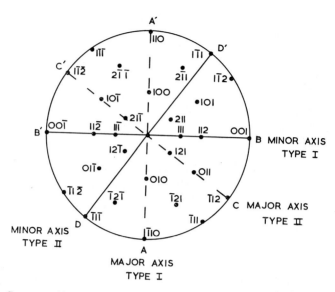

Fig. 3.6.   Stereographic projection of the major and minor axes of elliptical tungsten specimens showing streaks as in fig. 3.5.

not an interplanar spacing. In each instance a step is generated, of height a fraction of an interplanar spacing, which cannot be removed by field evaporation. Well characterised examples of such streaks are to be found in the work of Ralph (1964), Ranganathan et al. (1965) and Smith et al. (1968). Faulted dislocation loops, giving streak contrast can be generated by homogeneous nucleation under the high stresses operating during specimen preparation (Ranganathan et al., 1965; Smith and Bowkett, 1968b). Certainly faulted loops, again giving streaking, are nucleated in iridium field-ion tips, probably during field evaporation (Fortes and Ralph, 1968a). Significantly, the streaks in iridium and tungsten field-ion images lie on likely stacking fault planes.

In two phase systems the field evaporation characteristics of precipitate and matrix are seldom the same and it follows that there is likely to be a discontinuity in the smooth field evaporated end form at the particle-matrix interface which may give rise to image streaking.

### 3.2.3. Streaks having their origin in electronic effects

Holland (1963) postulated the existence of special electronic energy levels in association with defects such as stacking faults and grain boundaries.

The presence of such electronic states could locally increase the ionization probability and therefore the intensity of the field-ion image. Such an effect would account for the bright streak associated with the grain boundary shown in fig. 3.7. A similar effect may increase the brightness of streaks associated with stacking faults.

It will be appreciated that the origin of streak contrast in the field-ion image is not entirely clear and with Brandon (1966a) we urge extreme caution in the interpretation of image streaks except in symmetrical images. The occurrence of streaks can be limited by precautions in specimen preparation e.g. polishing back from the original fracture surface of the specimen, avoiding excessive resistive heating in the electrolyte and careful removal of polishing deposits by washing the specimen in appropriate solvents.

### 3.2.4. ZONE DECORATION

A conspicuous feature of field-ion micrographs from a number of metals is the presence of bright image spots along particular crystallographic zones of the image. This phenomenon is called zone decoration. Table 3.1 summarises the characteristics of the zone decoration on a variety of metal surfaces. Zone decoration occurs only along particular parts of a zone e.g. in micrographs of tungsten such as fig. 1.14 there are bright extra image spots on the [001] zone between the (130) pole and the (310) pole through the (110) pole. Zone decoration has not been observed on rhodium, palladium, cobalt or nickel. Since the zone decorating atoms are very bright it is likely that they occupy protruding low co-ordination number sites. Comparison of the image spot arrangement with the known atom arrangement in {210} planes of tungsten indicates that zone decorating atoms do not occupy lattice sites. (The {210} plane facets are on the decorated part of the ⟨001⟩ zones.) It should be noted then that information regarding image features on the ⟨001⟩ zones of tungsten micrographs is not reliable. Müller (1964) has interpreted the zone decoration phenomenon by postulating a contribution to the binding energy of an atom occupying a low co-ordination number site from polarization bonding. For example in bcc metals the polarization bonding contribution is sufficient to stabilise an atom in a metastable site with three near neighbours rather than four when:

$$\alpha_3 F_3^2 - \alpha_4 F_4^2 \geqslant \tfrac{1}{2}\Lambda .$$

$\alpha_i$ is the polarisability of an atom in a site of co-ordination number $i$, $F_i$ is the field at such a site and $\Lambda$ the sublimation energy.

The two observations that zone decoration occurs in specific image

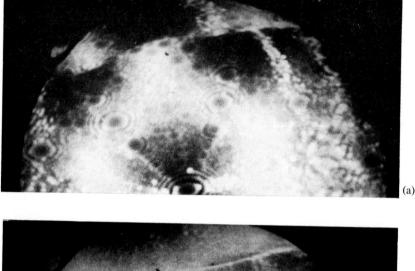

(a)

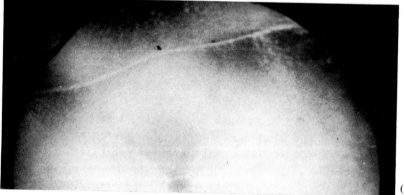

(b)

Fig. 3.7.   Streak associated with a grain boundary in a tungsten specimen. Figure 3.7b
indicates the effect of raising the voltage well above B.I.V.


regions and also occurs even in zone refined material make it unlikely that
the bright image spots are a consequence of impurities. The lack of detailed
knowledge of electronic states in the surface of a transition metal especially
in the presence of a large electric field means that at present it is possible
neither to predict the sites at which zone decoration will occur, nor to explain
why the details of the phenomenon are specific to each particular metal.
However, experiments using an 'Atom Probe' field-ion microscope by
Brenner and McKinney (1968) and Müller et al. (1968) indicate that zone
decoration atoms are of the same species as the matrix in iridium.

The appearance of a zone decorating atom is similar to that expected

TABLE 3.1

The characteristics of the zone decoration on a variety of metal surfaces (modified from Müller, 1967a)

| Metal | Decorated zone | Temperature of field evaporation | Comments |
|-------|---------------|----------------------------------|----------|
| W  | $\langle 001 \rangle$ | 4.2°K–700°K | |
| Ir | $\langle 110 \rangle$ | 60°K | |
| Pt | $\langle 110 \rangle$ | 60°K | |
| Re | $\langle 11\bar{2}1 \rangle$ | 4.2°K–700°K | |
| Mo | | 21°K–80°K | Diffuse Decoration |
| Ta | $\langle 110 \rangle$ | 200°K | |
| Fe | $\langle 001 \rangle$ | 80°K | With $H_2$ Promotion |
| Be | $\langle 0001 \rangle$ $\langle 12\bar{1}0 \rangle$ | No data | |

from an interstitial. It is relatively easy to distinguish most observations since zone decoration is localized. However Müller (1967a) and Brandon (1966a) have observed atoms in other metastable sites which in the case of niobium extend over the whole image at 21°K but were removed by field evaporation at 78°K. Similar metastable sites may occur above small electronegative interstitials, like oxygen for example, accounting for the unexpectedly large size and brightness of image spots associated with oxygen. It appears that it is not the oxygen itself which images in this way but rather a molecularly bonded metal atom occupying a metastable site above an oxygen atom (Cranstoun, 1968).

### 3.2.5. PSEUDO POINT DEFECTS

A vacant lattice site will show up in a field-ion image as a black spot where an atom spot might otherwise be expected to appear. The perhaps trivial possibility that a screen phosphor defect may give rise to this effect has already been mentioned, though an experienced microscopist would be unlikely to be taken in by it either because its shape, size and position are unlikely to correspond exactly to an atom site or because it remains stationary and unchanged during field evaporation. A more difficult case for distinction occurs at certain lattice sites that frequently appear dark, such as those where two series of plane rings intersect. The 'missing' atom is imaging only faintly on the screen, possibly because it is shielded by the atoms all round it and cannot receive a sufficient supply of imaging gas atoms. The faint image is then below the threshold light level for the photographic emulsion used in recording the image.

There remains the question of whether vacancies will be induced in a field-ion specimen by the field stress. This problem is considered in § 4.3.

### 3.2.6. DISLOCATION ARTEFACTS

The large value of the surface area to volume ratio of a field-ion tip has the consequence that pre-existing dislocation configurations may be changed by the operation of image forces. In addition the release of pinned dislocations by dissolution of obstacles to slip during tip preparation may result in a reduction in dislocation density of more than the 20% estimated by Hirsch et al. (1965) for thin foils. The configuration of a dislocation line is affected by image forces, e.g. a screw dislocation is constrained to emerge

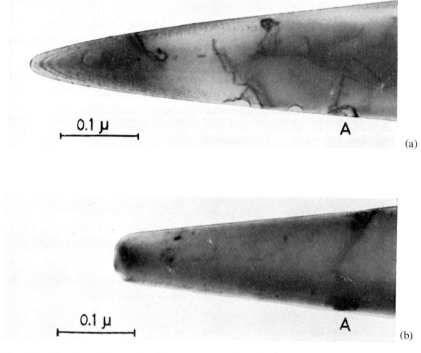

(a)

(b)

Fig. 3.8.    Electron micrograph of a tungsten specimen before (a) and after (b) field evaporation showing the tip region denuded of dislocations. Note, however, the considerable
dislocation density in the shank. (After B. Loberg and H. Nordén.)

normal to a free surface. In deducing values of stacking fault energy from the separation of partial dislocations some allowance can be made for the effect of the surface (Eshelby and Stroh, 1951).

The shear components of the applied field stress cause glissile dislocations to glide out of the tip as the field is raised for imaging. Electron microscopy has confirmed the presence of a zone denuded of dislocations in the region of the tip of a field evaporated specimen; fig. 3.8. A complicating factor when considering the origin of dislocations in field-ion tips is the capacity of the field stress to produce perfect or faulted dislocation loops by homogeneous nucleation, or the operation of a dislocation source. It has been observed that the field stress can also cause mechanical twinning in

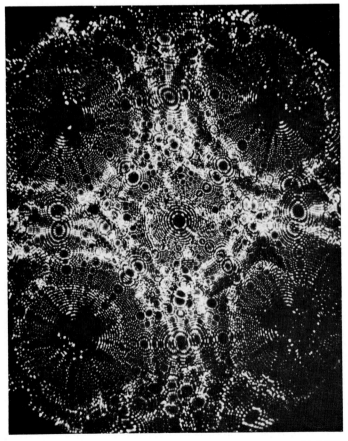

Fig. 3.9.　{111} slip traces in an image of platinum. (Courtesy E. W. Müller.)

specimens under observation. This type of deformation has been observed in iridium (Rendulic, 1966), platinum (Rendulic, 1967), tungsten (Bowkett, unpublished) and tungsten–5% rhenium (Ralph, 1964).

It is apparent from symmetry arguments that there is no resolved shear stress on an axial screw dislocation in an elastically isotropic field-ion tip. Since the yield stresses of bcc metals are strongly temperature dependent and increase as the temperature is lowered the observation of dislocations, which depends on the defect remaining in the specimen, is best attempted with the tip at 21 °K or less. It is common for slip steps to be produced on field-ion tips of materials with a low Peierls stress, e.g. cobalt or platinum (fig. 3.9). Since the deformation necessarily results in a lowering of the field such steps give a stable image as a family of approximately small circles with a common pole. In a manner corresponding to the removal of macro slip steps by chemical polishing, slip steps on a field-ion tip in principle could be entirely removed by field evaporation, However, raising the field for field evaporation may cause further slip to occur and preclude the attainment of a perfect field evaporated end-form.

Since the atoms at the core of a dislocation are not in lattice sites,

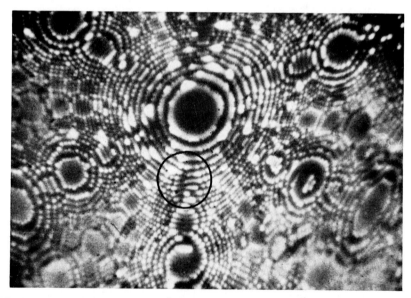

Fig. 3.10. Dislocation emerging in the (111) pole of an iridium specimen. Note the distortion of the rings adjacent to the beginning of the spiral. Dislocation spirals can also be seen in adjacent poles. (Courtesy T. F. Page.)

their binding energy is decreased. This effect may have the consequence of enabling field evaporation to occur at a reduced field, and would prevent the observation of core atoms by simple techniques and lead to an over-estimate of loop sizes. Similar considerations may apply to the atoms around vacancy clusters.

Where a dislocation spiral begins in one of the outer ledges of a field-ion pole there is an extra ledge, easily mistaken for an extra half plane as in the classical model for an edge dislocation. The extra ledge gives a local change in radius and alters the regular sequence of ledge widths (fig. 3.10). The discontinuity thus introduced is smoothed out by field evaporation resulting in an apparent bending of the lattice planes adjacent to the extra ledge (Fortes and Ralph, 1969a), this distortion is on a scale an order of magnitude greater than elastic deformation but may correspond to a small displacement normal to the specimen surface (see p. 87).

Müller (1958) argued convincingly that spiral structures could be produced by eccentric bunching of lattice planes during *in situ* annealing of field-ion tips. Such structures can be removed by field evaporation.

# 4 | POINT DEFECTS

## 4.1. Introduction

The field-ion microscope with its atomic resolution has an obvious application to the study of point defects. Indeed in many ways the microscope can be regarded as the only direct method of studying individual point defects, although there are other techniques which may be able to provide similar information and should not be ignored e.g. very high resolution electron microscopy and small angle scattering of electrons, X-rays or cold neutrons. Field-ion microscopy potentially has great advantages over such methods as the measurement of changes in resistivity, which may not be able to distinguish the *type* of point defects present in a specimen. A direct determination of the nature of the point defects (i.e. vacancy or interstitial) should be possible for any field-ion micrograph by simple inspection, and distinction between self-interstitials and impurity-interstitials may be possible with an 'Atom Probe' time-of-flight mass spectrometer (§ 2.7.4).

Examples of studies of point defects that can be made using field-ion microscopy are the 3-dimensional distribution of irradiation damage, the distribution of substitutional solute atoms, the interaction of interstitial atoms with dislocation lines and also the clustering of point defects around 1- and 2-dimensional defects.

## 4.2. Methods of introducing point defects

Substitutional and interstitial impurity atoms can be studied by using intentionally impure materials with a range of the impurity element. The image quality deteriorates markedly when the concentration of second element exceeds about 5%; but so long as the amount is kept below this, useful quantitative data may be obtained. Other methods of introducing foreign atoms are by ion implantation and by controlled vapour deposition followed by diffusion. Vacancies can be introduced by irradiation or in some cases by quenching, and vacancies may also be present in non-stoichiometric materials (see § 8.4.1). Self-interstitials can be introduced by irradiation.

For reactor irradiation experiments it is important to choose the total dose and energy of the neutrons so that within the volume of material examined in any one field-ion experiment there is a statistically good chance of finding a small number of primary events. In tungsten, for instance, at a total neutron dose of $10^{17}$ nvt about 1 atom in $10^6$ is displaced as a primary event, with a mean energy of 2.5 keV and a maximum energy of 32.5 keV. This primary then induces secondary displacements, and between 30 and 600 displacements are formed per primary. Self annealing in the displacement spike region might be expected to remove some of the damage, but the amount left is still considerable and readily detectable during field-ion examination. If irradiations to higher total doses are used overlap of displacement events might be expected to arise.

An attempt to correlate field-ion and electron microscope results on quenching of platinum has been made by Newman (1968) who found defects in the field-ion images which could be identified as faulted loops, although it was difficult to identify them positively as Frank (rather than Shockley) type. Fortes (1968) tried quenching iridium by resistance heating a wire so that it melted through and acted as its own switch giving quite a good quench. However, the treatment is so severe that it is not possible to prepare specimens from wire quenched in this manner which becomes very ragged and notched. A similar technique has been devised by Schultz (1964) and used by Galligan and Attardo (1966) to quench tungsten wires; in this case the specimen is immersed in a bath of liquid helium II at a temperature of about 2 °K and pulsed with a large current to the melting point. A very thin sheath of gaseous helium will form, around the heated specimen, which is reported to collapse rapidly upon quenching. The total time taken for heating and quenching is quoted as less than 0.05 sec and it is probably important that this should be a minimum in order to avoid the corrosion and pitting problems outlined above. A method (Meakin et al., 1964) that has been used with some success to introduce vacancies into thin foils of molybdenum, is to quench from a high temperature into a bath of molten indium, which melts at 156 °C i.e. below the reported temperature for vacancy migration in molybdenum. Therefore this method would probably be successful for quenching wires of bcc metals such as tungsten and molybdenum.

One other method of introducing point defects may be the field evaporation process. In a completely resolved plane, the field must be more or less uniformly high over every atom and it is difficult to predict the order in which the atoms will be field evaporated assuming that they are all in similar sites. For this and other reasons it may be dangerous to state even that a vacancy seen in the centre of a plane is a genuine one.

## 4.3. Imaging

There is an optimum number of defects for a field-ion specimen: enough to be certain of seeing some, without having appreciable interaction, or in the case of irradiation overlap of displacement spikes.

The ideal concentration of defects is probably about $10^{-3}$ which should give about 10 observations in each field of view assuming that something like 10000 atoms are seen in the field-ion image. The optimum number for subsequent interpretation depends on the method by which the defects are introduced: if they are produced by some event like a particle striking the specimen, it is simplest to avoid whenever possible an overlap of the effects of high energy particles. On the other hand after quenching, the upper limit to defect concentration is almost certain to be set by the quenching conditions rather than any consideration of field-ion imaging.

If there are less than 10 point defects in an average field of view there will not be an appreciable chance of detecting even one, on the basis that in most images the resolution is only good enough to detect point defects unambiguously over perhaps 10% of the image. Although 200 or more micrographs of a specimen may be taken the sample is still small and this will be reflected in the reliability of results such as defect concentrations. In the case of tungsten, for instance, imaged with helium ions at liquid nitrogen temperature the only regions of the image with a resolution that is consistently good enough to detect vacancies are often the edges of the {112} plane rings. Figure 4.1 shows examples of vacancy defects on the {112} ring edges in tungsten. Clear examples of vacancies on the {110} ring edges can also be found in tungsten, as in fig. 4.2, but do not commonly show up in micrographs taken with helium imaging at liquid nitrogen temperature, because the contrast of the {110} poles is too faint unless the field is raised and the remainder of the image blurred. The {100} poles in tungsten are

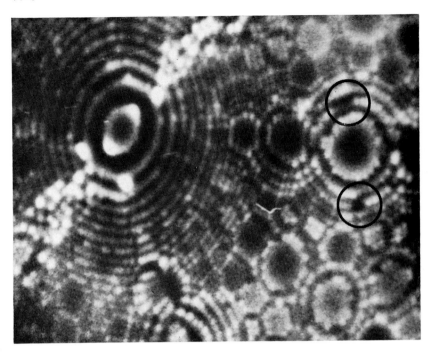

Fig. 4.1.  Vacancy defects on the {112} ring edges in tungsten.

usually ignored for vacancy counting purposes, because they field evaporate in a very uneven fashion and tend to show pseudo-clusters, as in fig. 4.3.

Misleading results giving vacancy concentration figures that are too high, can sometimes be traced to the counting of erroneous "vacancies" which appear in regions of the image that are either poorly resolved or of low contrast. For this reason the "observed" vacancy concentration can be sharply dependent on the imaging field, as was observed by Wald (1963), since each plane has its own individual best image voltage. In tungsten such difficulties are largely overcome by concentrating on the {112} regions of the image.

One problem in determining the contrast expected from vacancies arises from an uncertainty in the meaning of each image spot round a ring; it is possible that in some cases bright spots arise not from a single atom but unresolved rows of two or three atoms (Forbes, private communication). If this is the case an "atom" that appears not just white but as a shade of grey could correspond to (say) one vacancy in a row of three atoms. In

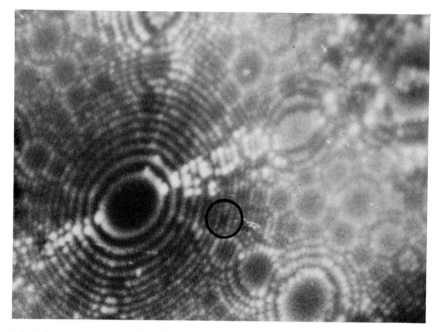

Fig. 4.2.   Vacancy on the {110} ring edges in tungsten; in general such vacancies only
show up clearly slightly above the best overall imaging field.

such a case, however, the darkening would not necessarily correspond to a
30% reduction in the ion current for the spot, since the supply function to the
remaining atoms could be increased giving an effect opposing the reduction
in ion-current. The situation is complicated because such a darkening
could also come from a vacancy just underneath the surface which might cause
a slight recession. Using a film with a sharp 'cut-off' for photographing the
image will tend to exaggerate the effect by showing up dim spots as black
or white, the black ones erroneously being counted as vacancies. This may
not matter so long as an absolute vacancy concentration is not required, but
it does emphasise the importance of keeping *all* the experimental conditions
constant if it is desired to compare the vacancy concentrations in two
specimens. The number of atoms making up an image spot probably varies
with the situation; in places where the image takes the form of a series of
round spots arranged in the expected crystallographic array (making
corrections for the projection and image distortions) each spot will almost
certainly arise from a single atom. However difficulties arise in cases like the

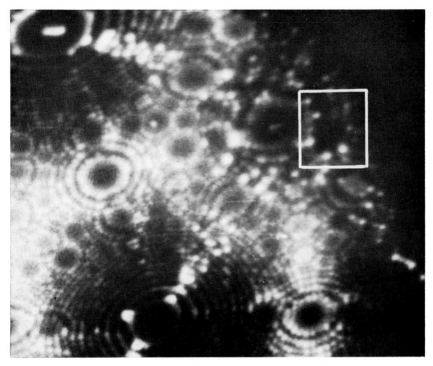

Fig. 4.3.  Pseudo-cluster in the {100} pole of tungsten, probably generated by uneven field evaporation.

{112} plane in tungsten where the close packed direction $\langle11\bar{1}\rangle$ has an atom spacing of 2.7 Å which may not be fully resolved, giving rise to eggshaped spots as in fig. 4.4. It is possible that the bright zone decoration spots arise from clusters of four or five atoms, although results obtained on iridium using time-of-flight mass spectrometers seem to indicate that in fact zone decoration spots arise from a single atom.

There remains the question of whether vacancies will be induced in a field-ion specimen by the field stress. The mechanical stress, $\sigma$, exerted on the surface by a field $F$ will be:

$$\sigma = F^2/8\pi. \tag{4.1}$$

For tungsten imaged with helium ions this stress amounts to about $10^{11}$ dyne cm$^{-2}$. As a working hypothesis the stress can be thought of as a negative hydrostatic pressure, in which case a volume expansion of about 10% might be expected from the known values for bulk modulus, and the

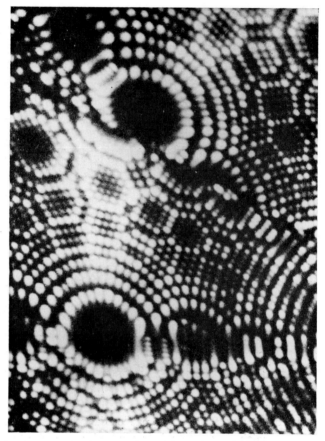

Fig. 4.4. {112} planes in tungsten; the egg-shaped spots may arise because the atomic chains in the $\langle 11\bar{1} \rangle$ direction (spacing 2.7 Å) are not fully resolved. (Courtesy J. J. Hren.)

question arises as to whether vacancies will be formed spontaneously under such conditions. Following Müller (1965) we may consider the case of vacancies in tungsten, by considering the volume $V_a$ of a vacancy to be equal to the atomic volume $15.8 \times 10^{-24}$ cm$^3$; the energy of increasing the volume of a tip is then:

$$p \, dV = (F^2/8\pi) \cdot V_a = 2.0 \text{ eV} \quad \text{at an evaporation field of 6.8 V/Å.}$$

This energy can be compared with the reported values for the energy of formation of a vacancy in tungsten of at least 3 eV (Brooks, 1955; Kraftmakher and Strelkov, 1963). However, it is difficult to relate the energy calculated for bulk vacancies to the situation on the surface of a field-ion specimen (though the energies might be expected to be in an approximate ratio of 3:2), and it is essential to make an experimental check by carrying out a vacancy count on unirradiated material. It was found by Bowkett and Ralph (1969) that large numbers of vacancies are not induced in tungsten specimens by the field. In the case of iridium and platinum however, there is some evidence that vacancies may be induced at evaporation fields (see § 4.5.3).

The possibility of preferential field evaporation of atoms from the edges of vacancy clusters is a more complicated problem. It is clear that the atoms adjacent to a cluster will have a reduced co-ordination number which means that they will either be preferentially evaporated (Brandon, 1964c) or be stabilised by field penetration (Müller 1964). Practical observation suggests that the atoms surrounding the cluster will in fact be stabilised and in consequence, clusters appear in the image with their correct size.

Interstitial atoms show up in the image as additional bright atoms in unexpected positions. Müller (1959) suggested that the bright contrast arose not from the interstitial atoms themselves but from interstitial atoms just beneath the surface pushing out the otherwise flat surface. Computer simulation of images (appendix 1) would appear to support this idea. On the basis of the Moore "thin-shell" model (Moore, 1962) we know that the image is very sensitive to displacements along the local normal, and a bulging of even 0.05 of the lattice parameter could easily lift an atom into or out of the imaging shell (since the shell thickness is typically 0.1a). Clearly a very small bulge may give a visible effect.

There should be little difference in contrast between self-interstitial atoms and solute interstitials, both of which will give rise to surface bulges. A *substitutional* solute atom could also give rise to a small surface bulge, or collapse, if it is larger or smaller than the matrix atoms, and give rise to bright or dark contrast. All this assumes that the computer shell model has an application to the real criterion for imaging and that a normal displacement of 0.05 $a$ (or probably even less) affects the imaging.

A distinction between substitutional and interstitial atoms giving rise to similar contrast could be attempted using time-of-flight mass spectrometry experiments although the interpretation of the results may not always be easy. For instance consider an alloy AB (e.g. Pt-Au) where the B atoms in general do not appear in the field-ion image because either they are preferentially evaporated or because they simply do not image. In "pure" metal A a bright spot on the surface could be composed of A atoms even though it might arise from a surface bulge due to a subsurface B impurity atom. It would be all too easy to interpret this wrongly as a self-interstitial unless the evaporation were continued far enough to detect the B atom. There is also the other case where the foreign atom may be preferentially retained e.g. Re in W (Elvin, 1967), which corresponds to the case of A solute in B rather than B solute in A as before. Here a bright spot could either arise from a retained impurity (A) or a surface bulge (B) due to a subsurface impurity (A). If examples of these two cases could be found it would be very interesting to compare the imaging characteristics with the mass spectrometry results.

A useful idea would be to develop a technique where it is possible to collect and analyse *all* the evaporated material (e.g. as being 84% of tungsten, 16% of rhenium) and compare the results with the observed number of bright spots in the corresponding images.

### 4.4. Analysis of results

For the reasons outlined in the above section, it is unlikely that the field-ion microscope can be used readily to measure *absolute* defect concentrations. However, it can be used with advantage to follow changes in concentration after various treatments. For instance for most metals vacancy concentrations are obtained that contain an element that arises from image artefacts or counting errors or even from vacancies introduced by the evaporation field. If, however, counts are made under the same conditions both on irradiated and unirradiated material, the *difference* in concentration will bear some relation to the effects of the irradiation. Note that even after subtracting the 'spurious' vacancy concentration, a measured concentration may still not be genuine, since for instance not all vacancies may have been located. There is also a problem in deciding on a suitable number of observations to make in order to get statistically significant results. In general the maximum number possible should be analysed, but the analysis can be a tedious business and it is difficult to devise a method by which some sort of automatic scanner could recognise a point defect. Care must be taken to avoid confusing genuine bulk vacancies with image artefacts due to impurity atoms which give dark contrast: this is best done by making a statistical analysis.

There are two simple ways in which the data gathered with a field-ion microscope can be analysed. It is common to record the screen image on 35 mm film, and the film can be projected using for instance either a film-strip projector or a micro-film reader. There are two important disadvantages to this method of analysis, however; the film is liable to be damaged either by heat in the projector or by scratching, and there is no permanent record of the results, which can be referred to for checking. Alternatively, a positive enlargment can be made from each negative. This provides a permanent record, but involves a considerable amount of additional work and expense. In addition the high level of contrast inherent in a field-ion micrograph means that it is very difficult to make prints without losing some information in the process (this is especially true where large numbers of prints must be made).

It has been found that vacancy counting can be carried out most successfully on positive enlargement prints, and the accuracy depends strongly on the size of the enlargement; prints about 20–25 cm square being the optimum size.

The information that can be obtained by projection of the micrographs is of a different kind; projection enables a rapid examination of all the micrographs in a sequence, so that trends in the distribution of the damage can be spotted. This is conveniently done by transferring the negatives to a ciné-film (or, better, recording the information on ciné-film initially) which can be examined at varying speeds in an analysing projector.

The problem of vacancy counting is partly a psychological one. Looking at any random micrograph from an irradiated material an observer is mentally predisposed to find four or five defects (e.g. vacancies) on it and once the four of five have been found it becomes difficult to find any more. If there are no defects, there is always a chance of marking spurious effects, no matter how hard one tries to prevent this. At least actually marking the defects on the photographs makes it possible to come back later and check the identification.

Photographing the image after the evaporation of individual atomic layers does not give sufficient information to distinguish between single and divacancies or to find the exact size of small clusters. Since only protruding atoms are imaged, it is necessary to do the evaporations on an atomic scale in order to record every atom site. Figure 4.5 may make this point clearer. In practice, 8–10 photographs per (110) plane removed are found to be a sufficient number for every atom to be examined. The evaporation is controlled either by observing the image through a pair of X10 binoculars whilst the field is raised towards evaporation voltage, and lowering the field again as soon as a few atoms have been removed, or *far more accurately* by using a pulse generator to pulse the high tension supply to the specimen a predetermined amount, which is sufficient to evaporate about a tenth of an atomic layer from the surface.

Fig. 4.5.   The amount of material removed when "one plane" is evaporated. In order to
image every atom site it is necessary to do finer scale evaporation.

The field-ion microscope, used in this way, gives a complete 3-dimensional picture of fine damage, but against this must be weighed the disadvantage that the gathering and especially the analysis of the data is exceedingly tedious. Although the specimen surface will remain entirely clean and uncontaminated so long as the field is applied (since no impurity molecule will be able to approach the tip), it will contaminate immediately the field is removed, at a rate dependent on the background pressure of impurities in the system. This means that it is often necessary to take a large number of photographs without leaving the microscope for more than short periods. Even using some method of image intensification to reduce the photographic exposure time, the actual time spent analysing the micrographs will still be appreciable.

## 4.5. Experimental results

### 4.5.1. INTERSTITIAL IMPURITIES

Müller (1959) noticed that field-ion specimens of rhodium prepared from commercially pure (better than 99.9 per cent) wire showed a number of bright randomly situated spots with a concentration of $3 \times 10^{-4}$ of the rhodium atoms. Heating the specimen tips up to about 600 °C in vacuum did not remove the bright spots and Müller proposed that they arose from oxygen atoms in interstitial sites. He also reported similar observations made on 99.99 per cent pure platinum tips where the bright spots could be introduced by annealing the specimen in air or in a low pressure ($10^{-4}$ Torr) of oxygen and they could be removed by annealing the tip in the $10^{-6}$ Torr vacuum of the microscope.

Machlin (1967) has shown that interstitial oxygen atoms in tungsten give rise to similar bright spots. He heat treated tungsten wires so as to vary their oxygen content and found that tips made from these wires showed a systematic variation in the number of bright spots seen with the oxygen content. An attempt to repeat the experiment with wires treated to contain carbon instead of oxygen gave rise to very low bright-spot counts which indicates that either the carbon solubility limit at 2000 °C is less than 10 ppm or, more probably, that carbon is not visible as an interstitial solute in tungsten. In tungsten it is necessary to restrict bright spot counts to regions well away from the [001] zone decoration line, and Machlin restricted his counts to triangular regions bounded by the edges of the (110) and two adjacent {112} poles as shown in fig. 4.6. Machlin also

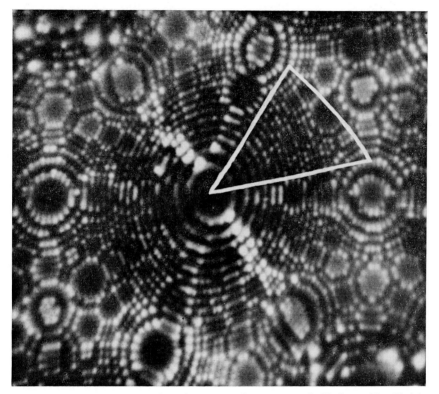

Fig. 4.6.   The approximate region of a tungsten image most suitable for making "bright spot" counts.

drew attention to the changes in intensity, size and shape of a bright spot that occur depending on its position relative to a ring ledge, as illustrated in fig. 4.7: the bright spot becomes larger and brighter as the ring ledge sweeps in towards it during evaporation. The shape distortion of the bright spot is typical, but the reason for it is not known.

Bright spots in iridium specimens have been attributed to interstitial

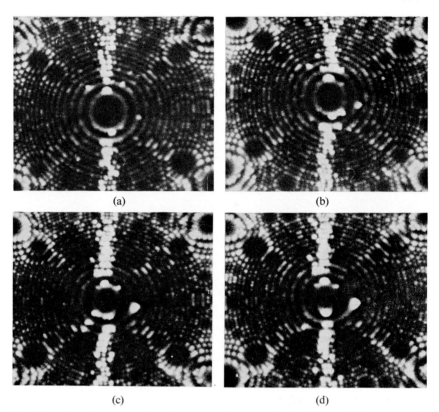

(a)                                    (b)

(c)                                    (d)

Fig. 4.7.   The changes in intensity, size, and shape of a bright spot that occur as its
position changes relative to a ring edge. (Courtesy E. S. Machlin.)

oxygen by Ranganathan (1964), and Fortes and Ralph (1967) have studied
their distribution and field evaporation behaviour. They counted the bright
spots in the {111} regions (i.e. away from the zone decoration in the {001}
regions) of about 40 (111) planes in each of two iridium tips. The oxygen
atom concentration measured in this way was $4.7 \times 10^{-4}$ of the iridium
atoms i.e. 470 ppm of oxygen, which compared well with the manufacturers
figure of 500 ppm. Fortes and Ralph (1967) also studied the segregation of
oxygen (bright atoms) to grain boundaries in iridium.

Nakamura and Müller (1965) have investigated the initial stages of the
oxidation of tantalum, and have shown examples of the early stages of the
precipitation of oxygen from the matrix when groups of two or three oxygen
atoms begin to cluster together. Further relevant work is described in ch. 8,
and some experimental work on *substitutional* alloys is described in ch. 9.

4.5.2. SELF-INTERSTITIALS (see also § 8.2)

Interstitial metal atoms of the same kind as the matrix can be produced by bombarding the tip with particles of various energies. Self-interstitials may give rise to the same type of contrast as impurity interstitials. Experimentally the simplest type of bombardment is that produced by reversing the imaging field so that the gas atoms that would otherwise be used for imaging the specimen are drawn in and strike the specimen transferring an energy of up to 500 eV. Sinha and Müller (1964) have bombarded tungsten tips with a well-collimated beam of 20 keV helium atoms and both Müller (1959) and Brandon et al. (1961) have bombarded specimens with α-particles (see § 8.2). In all these experiments a number of vacancies and bright interstitial spots appeared at the surface.

The interpretation of bright spots as displaced atoms in interstitial sites is supported by the damage annealing experiments of Sinha and Müller (1964). If a tip is bombarded with He atoms at 21 °K bright spots continue to arrive at the screen for a short time after the bombardment has been stopped which Sinha and Müller attributed to shallow lying interstitials diffusing to the surface with a lowered activation energy for diffusion due to the field stress. As the bombarded tips were warmed up it was found that displaced atoms diffused to the surface most rapidly in the temperature range 85–95 °K. Similar results were obtained after bombardment with individual α-particles from a $^{210}$Po source. Petroff (1967) bombarded iridium at about 5 °K with 10 MeV protons and found during post-irradiation heating extra bright spots appeared at the surface in the temperature range 18 °K–40 °K. Further increase from 40 °K to 160 °K did not show any new bright spots on the surface, although there was a certain amount of surface re-arrangement of the {111} and {113} planes in the temperature range 100 °K to 160 °K. On the other hand, Attardo et al. (1967) have examined tungsten that had been reactor irradiated to a high dose (between $5 \times 10^{19}$ and $2 \times 10^{20}$ fast neutrons/cm$^2$) and have measured an interstitial (i.e. bright spot) concentration between $10^{-3}$ or $10^{-5}$ after irradiation and less than $10^{-5}$ after postirradiation annealing. These results were interpreted as arising from interstitial removal in stage III recovery and do not appear to agree with the above observations or with results obtained by Bowkett and Ralph (1969) (see § 4.5.3).

4.5.3. VACANCIES (see also § 8.2)

Vacancies introduced by quenching have been studied in platinum by Müller (1959) who measured a vacancy concentration of $5.9 \times 10^{-4}$ in a specimen that had been quenched from 1800 °K. The concentration was measured by counting the number of vacancies seen on the (102) plane. Pimbley et al. (1966), however, have reported the examination of a number of specimens of annealed platinum which showed apparent vacancy concentrations in the range 1 to $20 \times 10^{-3}$, whereas the concentration expected in such material is much less. They interpret their results as showing that vacancies in the {102} planes of platinum are artefacts caused by the field stress during field evaporation. In view of the arguments presented in § 4.3 this is not unreasonable. Furthermore it is probable that the extent of this effect would vary from specimen to specimen and even from region to region of a single specimen, which is supported by the observation in the work of Pimbley et al. that the {023} and {531} planes of platinum showed a considerably smaller number of vacancies.

It is clear that great care must be taken when investigating vacancies only to work with materials that are not damaged by the stress arising from the evaporation field. Bowkett and Ralph (1969) have reported that few, if any, vacancies are created in tungsten by the imaging stress and Attardo and Galligan (1966) have reported that the observed fraction of vacant sites in annealed tungsten is considerably less than $10^{-5}$, which means that tungsten at least can be used for investigations of single vacancies. At the other extreme it is necessary to be certain that the resolution of the technique is sufficient to detect point defects: thus Rao and Thomas (1967) who looked at the {110} planes of irradiated molybdenum tips at 78 °K and reported finding *no* vacancies at all, were probably measuring the resolution of their molybdenum images rather than the vacancy concentration.

As an example of the manner in which the results of irradiation experiments can be analysed, we shall consider in some detail the analysis of some vacancy counting data from tungsten specimens that had been irradiated and then subjected to certain post-irradiation annealing treatments (Bowkett and Ralph, 1969).

It was found that vacancy counts had to be made in regions well away from grain boundaries which played an important part in deciding the final damage pattern. Vacancy clusters were found close to grain boundaries; fig. 4.8 is an example. This particular cluster occurred 15 Å from the high-angle grain boundary and was about 10 Å diameter; corresponding to some 50 vacancies. It is possible that it may have been formed as a result of the interaction of a channelon with the grain boundary. The interaction of a focussed collision sequence with the boundary is an unlikely possibility,

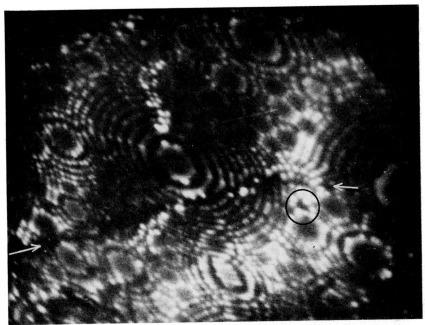

Fig. 4.8.　Vacancy cluster in tungsten about 15 Å from a high-angle grain boundary.

since the maximum energy that can be translated in a focusson is probably not sufficient to create a damage region this large.

Fortunately the approximately stereographic nature of the field-ion image means that the misorientation of the grain boundary can readily be found (see § 6.2.1); in this case the lattice rotation is ~ 52° about [110]. In the cubic lattice $\langle 112 \rangle$ directions lie in the (110) plane at angles to the $\langle 110 \rangle$ direction of 54° 44′. It is almost certain, therefore, that a $\langle 110 \rangle$ channelon would become a $\langle 112 \rangle$ channelon after crossing the grain boundary; the 2–3° deflexion, however, would be sufficient for the channelon to lose a large part of its energy to the lattice in a damage cascade close to the grain boundary. A possible objection to this explanation however, is that the energy necessary to cause a damage region of the size found is about 2.5 keV which suggests that the channelon had an unusually high energy. An alternative explanation is that the damage could be the debris of a primary event, though the proximity of this type of damage to grain boundaries is puzzling since a grain boundary should present no particular barrier to a neutron, and chances of random neutron events occurring close to a boundary by coincidence are remote.

In order to plot out the damage spectra, ideally every micrograph should be closely examined and the exact shape of any vacancy clusters, accurately drawn out. Such a detailed analysis is not possible on the many thousands of micrographs necessary for the determination of vacancy concentrations. Even if the time were available it might not be a desirable method of analysing the results, since it would be difficult to maintain a constant standard of accuracy.

A technique which can be used for the determination of the vacancy concentration from micrographs which is both simple and reproducible, but suffers from the drawback that it wastes some of the information present in the micrographs is as follows. Large photo prints should be used and all vacancy observations marked on them, be they single vacancies or small

(a)            (b)

(c)            (d)

Fig. 4.9. Representation using a ball-model of a section through a field-ion tip containing a vacancy cluster. As field-evaporation proceeds (a)–(b)–(c) the apparent size of the cluster will vary (see text). (d) is a plan view of the specimen. For simplicity the case of a cluster emerging on (110) is considered.

clusters. As discussed in § 4.3, in tungsten this analysis is often best confined
to the {112} plane edges. Successive micrographs between which small
amounts of field evaporation have been done can then be compared to check
whether a vacancy appears in more than one photograph – if it does *not*
the "cluster" size is noted down as $1/x$, where $x$ is the number of evaporations
for the removal of the particular (110) plane ring*. However, if the vacancy
appears on more than one micrograph the cluster diameter is $n/x$ where $n$
is the number of successive micrographs on which the vacancy or cluster
appears. Consider fig. 4.9c. During the evaporation of the plane indicated
the cluster will have an apparent maximum diameter of four interatomic
distances, but unless it were a disc the cluster would also appear during
evaporation on previous and succeeding planes. Suppose that in fig. 4.9c
the top plane were removed in a total of 12 steps: the cluster persistence
would then be $n/x = 8/12 = 2/3$ which corresponds to a diameter of 2/3 of
the ledge width i.e., about 11 Å diameter (say 50 vacancies) if the cluster
persists for less than one plane evaporation, and must be used in conjunction
with individual inspection for assessing the size of large clusters in order
to avoid confusing one cluster with a family of discs. The number of evaporations per (110) plane removed should be kept close to the optimum value
of about twice the number of atoms in a ledge width, to ensure that all
atom sites are examined. The method of analysis assumed that the small
clusters are spherical and that equal amounts are evaporated between each
successive micrograph during the evaporation of one plane ring. Regular
amounts of field evaporation can be achieved most easily using a pulse
generator as mentioned earlier.

An example of a field-evaporation sequence through a cluster in iridium
emerging in the (113) plane which persisted while a shell of thickness about
17 Å was evaporated, is shown in fig. 4.10.

The data obtained from vacancy counts on the irradiated tungsten
specimens we are considering is summarised in table 4.1. In this table the
abbreviations used mean that specimens were prepared from material in the
following conditions:

A – unirradiated tungsten.

B – as irradiated, $10^{17}$ fast $n \cdot cm^{-2}$.

C – irradiated $10^{17}$ fast $n \cdot cm^{-2}$ and annealed 400 °C, 3 hrs.

The results for specimens B and C are also presented in the form of

---

* As discussed earlier, due to image blurring, field evaporation is best controlled by
following the (110) pole.

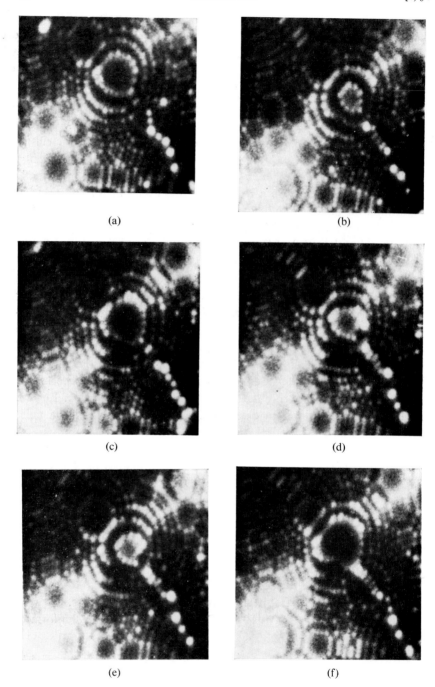

(a)

(b)

(c)

(d)

(e)

(f)

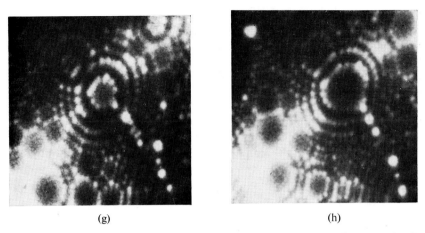

(g)                                                    (h)

Fig. 4.10. (a—h).   Field evaporation sequence through a cluster in iridium emerging in the (113) plane. The void is not present in the first or last pictures of the sequence. (Courtesy J. A. Hudson.)

histograms (fig. 4.11). In the histograms the block width is 0.2 of an evaporated (110) layer: this is about the size of a single vacancy (see later). It would be unrealistic to plot the figure with a block width of 0.1 of a (110) layer unless it were possible consistently to make over 10 evaporations for the removal of each individual (110) layer. This criterion is not met by the present results.

The results presented in table 4.1 and fig. 4.11 emphasize the very small volume of material that is examined in a field-ion microscope specimen. The volume is approximately 350 Å × 350 Å × 125 Å. Because local variations in vacancy concentration are to be expected until the neutron dose is high enough to give saturation effects, it is not permissible to make a direct comparison between the vacancy concentrations in the various specimens. Instead, it is necessary to compare the *ratios* of single vacancies to divacancies, clusters, etc. and follow the changes in the ratios, to determine the mechanisms of radiation annealing. For instance, if single vacancies are clustering, we can expect a decrease in the ratio of single vacancies to divacancies (perhaps accompanied by an increase in the ratio of large clusters to small clusters).

The results are presented in the form of ratios in table 4.3. Before determining the ratios, the vacancy concentration found in unirradiated material has been deducted from both of the irradiated concentrations. We may suspect that most of the "vacancies" in the unirradiated specimen

TABLE 4.1

Number of vacancies in size ranges for irradiated tungsten specimens after various heat treatments (the figures under the headings $A'$, $B'$ and $C'$ are the figures predicted for the evaporation of one hundred (110) planes)

| Size ranges: (units explained in text) | $A$ | $B$ | $C$ | $A'$ | $B'$ | $C'$ |
|---|---|---|---|---|---|---|
| 0 –0.1 | 15 | 126 | 35 | 22.4 | 274 | 49.5 |
| 0.11–0.2 | 62 | 64 | 140 | 92.5 | 139 | 197.0 |
| 0.21–0.3 | 21 | 14 | 32 | 31.4 | 30.5 | 45.0 |
| 0.31–0.4 | 8 | 6 | 43 | 11.9 | 13.0 | 60.7 |
| 0.41–0.5 | 8 | 9 | 13 | 11.9 | 19.5 | 18.3 |
| 0.51–0.6 | 2 | 0 | 7 | 3.0 | 0 | 9.9 |
| 0.61–0.7 | 1 | 0 | 3 | 1.5 | 0 | 4.2 |
| 0.71–0.8 | 1 | 0 | 2 | 1.5 | 0 | 2.8 |
| 0.81–0.9 | 0 | 1 | 0 | 0 | 2.2 | 0 |
| 0.91–1.0 | 1 | 3 | 4 | 1.5 | 6.5 | 5.6 |
| $> 1.0$ | 0 | 1 | 1 | 0 | (2.2) | (1.4) |
| No. of (110) planes evaporated | 67 | 46 | 71 | 100 | 100 | 100 |

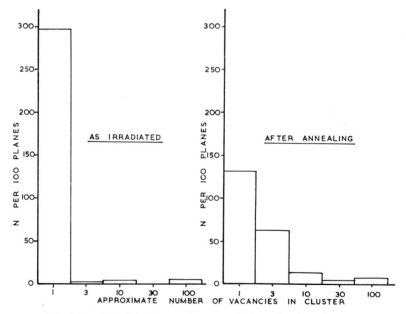

Fig. 4.11.    The distribution of vacancy clusters of various size ranges in irradiated tungsten specimens.

arose from counting errors, or are in any case artefacts peculiar to the field-ion microscope, but in this case the "observed" vacancy concentrations in each size range will be typical of that to be found in *any* specimen, irradiated or otherwise, so that it is justifiable to make a straight subtraction. The corrected concentrations are those given in table 4.2.

It is now necessary to convert the cluster sizes into a more recognisable form. The field evaporation analysis measured the cluster diameter and it is assumed that all the vacancy clusters are spherically symmetrical. It is acknowledged that this is an approximation; however, a number of spot checks which have been made, have confirmed the validity of the assumption.

TABLE 4.2

The figures of table 4.1 corrected by subtracting the vacancy concentrations in unirradiated specimen A

| Size ranges: (units explained in text) | B | C |
|---|---|---|
| 0 –0.2 | 298 | 132 |
| 0.21–0.4 | 0.2 | 62.4 |
| 0.41–0.6 | 4.6 | 13.3 |
| 0.61–0.8 | (− 3.0) | (4.0) |
| 0.81–1.0 | (7.2) | (5.6) |
| > 1.0 | (2.2) | (1.4) |

TABLE 4.3

Vacancy and vacancy cluster concentrations presented in the form of size ratios

| | | B (as-irradiated) | C (irradiated and annealed 400°C, 3 hrs) |
|---|---|---|---|
| I | $\dfrac{0 \ -0.2}{0.2-0.4}$ | 1490 | 2.1 |
| II | $\dfrac{0 \ -0.2}{0.2-1.0}$ | 33.3 | 1.6 |
| III | $\dfrac{0 \ -0.4}{0.4-1.0}$ | 33.8 | 8.5 |

I is roughly the ratio of single vacancies to divacancies.
II is roughly the ratio of single vacancies to divacancies and clusters.
III is roughly the ratio of single vacancies and divacancies to clusters.

Table 4.4 lists the estimated cluster sizes corresponding to vacancy persistences measured from the micrographs. Two different methods are adopted for relating persistence data to size measurements: when the persistence is between 0 and 0.7 it can be related to a *width within* a {112} layer. When the persistence is greater than this value it implies that the *depth* of the defect is greater than one {112} spacing and spurious vacancies (in fact part of the cluster) will be seen on *vertically* adjacent planes. In such cases the cluster *size* is better determined by individual inspection; few examples were found in this particular work. In order to convert persistences into cluster sizes conversion factors must be estimated for the particular plane and tip radius; for the present specimens the tip radii were all about 400 Å and the conversion factors are given in table 4.4. Assuming a sample volume of $6 \times 10^{-17}$ cm$^3$, the *total* vacancy concentrations of single vacancies and divacancies seen in the irradiated specimens were:

B (as-irradiated)                 $8.0 \times 10^{-5}$
C (annealed 400°C, 3 hrs)   $8.5 \times 10^{-5}$.

TABLE 4.4

Conversion factors for cluster sizes measured from persistence data

| Size ranges: (units explained in text) | Number of vacancies | | |
|---|---|---|---|
| | Minimum | Maximum | Weighted average |
| 0 –0.2 | 1 | 2 | 1 |
| 0.2–0.4 | 2 | 6 | 3 |
| 0.4–0.6 | 5 | 18 | $\approx 10$ |
| 0.6–0.8 | 10 | 50 | $\approx 30$ |
| 0.8–1.0 | 20 | 200 | $\approx 100$ |

As stressed earlier, field-ion microscopy is *not* a good method for determining vacancy concentrations and the above figures may have no *absolute* significance. The results suggest that single vacancies have migrated to small clusters after stage III annealing (400°C). Considering the ratio of single vacancies to divacancies (effectively line I of table 4.3), it is clear that the damage, which consisted predominantly of single vacancies after irradiation, has changed after annealing for 3 hrs at 400°C so that roughly equal numbers of vacancies are present in the form of single vacancies and divacancies.

Clusters containing more than about 100 vacancies were not detected in statistically significant quantities, although it is possible that if any very large clusters were present they caused the specimen to fracture before it could be imaged. It proved extremely difficult to image specimens after stage IV annealing (700 °C, 6 hrs); they tended to show a large number of dislocations but few, if any, single vacancies or small clusters, and fractured under the field stress.

Accuracy of vacancy counts can be checked as it was in this work by dividing the data from each specimen into four blocks, rather than taking it all together as previously, and determining the ratio of single vacancies to divacancies for each block of data. The scatter for any one specimen was less than 10%, provided all the measurements on the micrographs were made by one person. However, the scatter could rise to nearly 50% when comparing counts made by different people. It appears that only a fraction of the vacancies present are found and some observers achieve a higher score than others. Clearly, it is difficult to argue that vacancy concentrations measured from field-ion micrographs represent absolute values, but comparisons between counts made under identical conditions can allow changes in vacancy distribution to be followed.

# 5 | LINE DEFECTS

## 5.1. Introduction

One of the remaining major problems of dislocation theory is to elucidate the arrangement of the atoms at the core. Whether this is interpreted as the definition of the coordinates of the core atoms or the detection of very fine scale dislocation dissociations, the problem is below the routine resolution of the transmission electron microscope. The field-ion microscope offers a further degree of resolution and a number of observations have been made of dislocation cores in the latter of the senses mentioned above. In order to describe these observations we must first develop a dislocation contrast theory for the field-ion microscope.

## 5.2.  Single perfect dislocations

### 5.2.1. CONTRAST

Cottrell (1964) pointed out that *any* perfect dislocation line in a direction unit vector *l*, and having Burgers vector *b*, converted a stack of atomic planes, having unit normal *n*, into a helical ramp. Although this property is obvious for a screw dislocation, it is, in fact, common to all dislocations providing that $n \cdot b$ and $n \cdot l \neq 0$. Pashley (1965), Ranganathan et al. (1965) and Ranganathan (1966a) realised that such a helical ramp would give a spiral image in a field-ion micrograph.

One pole of a field-ion tip, when perfect, is sketched in fig. 5.1a. Locally the surface is flat and characterised by unit normal *n*. Since the atoms which are imaged are those within about 0.1 *a* (appendix 1) where *a* is the lattice para-

104

meter of a smoothly curved envelope (Moore, 1962) the projected image of the surface will be a series of concentric rings. The Burgers vector of a dislocation is the closure failure of a Burgers circuit (Frank, 1951). With the prescription that the circuit always passes through "good crystal", the Burgers vector is independent of how the Burgers circuit is drawn. The circuit can be drawn in the plane of unit normal **n** without loss of generality.

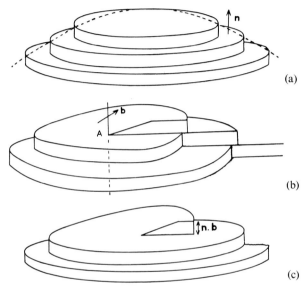

Fig. 5.1. (a) Sketch of one pole of a field-ion tip when perfect. The dashed line shows the smoothly curved envelope. (b) As (a) but after cutting and displacing an amount b, thus introducing a dislocation at A. (c) As (b) after field evaporation.

Any circle drawn in this plane and enclosing the dislocation is converted into one turn of a helix of pitch equal to **n·b**. In any lattice, with unit cell edges, **a**, **b**, and **c** the lattice vector $u\mathbf{a} + v\mathbf{b} + w\mathbf{c}$ from the origin terminates on the $p$ th plane of the set with Miller indices ($hkl$) where*:

$$hu + kv + lw = p.\tag{5.1}$$

We define an equivalent quantity $q$ to characterise partial dislocations:

* The parameter $\mathbf{g}\cdot\mathbf{b}$ where **b** is the dislocation vector and $\mathbf{g} = h\mathbf{a}^* + k\mathbf{b}^* + l\mathbf{c}^*$ where $\mathbf{a}^*$, $\mathbf{b}^*$ and $\mathbf{c}^*$ are reciprocal lattice vectors is another way of arriving at $p$ values and is preferred by some authors. It is of course identical with $p$ defined above and has the attraction of being equivalent to the electron microscopists' simple criterion for visibility of dislocations. We hope that it will be clear in what follows from the context whether **b** stands for Burgers vector or the cell edge.

eqs. (5.2.) and (5.3). Provided $n$ is normal to a rational lattice plane, and since a perfect dislocation has a Burgers vector $b$ equal to a lattice vector, it follows that the product $n \cdot b$ is always equal to an integral number of interplanar spacings. In the same way since $h$, $k$ and $l$, $u$, $v$ and $w$ are all integers $p$ is also an integer. This results holds for any lattice, e.g., in bcc $h + k + l$ is always even and with $b$ of the type $\frac{1}{2}a\langle 111 \rangle$ $p$ is integral etc. Similarly for fcc where $h + k + l$ must be unmixed and $b$ is of the type $\frac{1}{2}a\langle 110 \rangle$, $p$ is of the form $\frac{1}{2}(h + k)$ which again is always an integer. Figures 5.1b and 5.1c illustrate how a perfect dislocation intersecting a field-ion tip produces a spiral configuration. In fcc and bcc metals, where all the atoms are at lattice points, a perfect dislocation intersecting such a field-ion tip in a pole ($hkl$) always produces a spiral (except when $n \cdot b = 0$) which is of pitch equal to an integral number of ($hkl$) spacings. Computer simulation also predicts spiral contrast when a perfect dislocation intersects a field-ion tip (Ranganathan, Sanwald and Hren, 1966, Brandon and Perry, 1967).

 Equation (5.1) would also apply to crystals with more than one atom per lattice point provided that the indices $hkl$ always refer to the imaged atom planes, e.g., fully ordered cobalt-platinum and the hexagonal metals. Clearly, if $p = 1$ a simple spiral of pitch equal to the interplanar spacing is expected. When $p > 1$ the pitch of the spiral is equal to a number of planar spacings and field evaporation leads to the development of $p$ interleaved spirals (fig. 5.2).

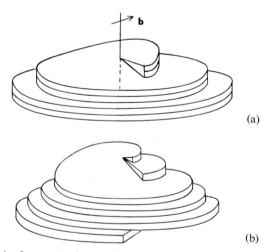

(a)

(b)

Fig. 5.2. (a) Sketch of one pole of a field-ion tip intersected by a dislocation with $p = 2$, before field evaporation. (b) As (a), but after some field evaporation which has reduced the enhanced field at the step of height $2d_{hkl}$ shown in (a).

When $p=0$, no spiral is expected, but the dislocation may give rise to contrast as a consequence of the different field evaporation behaviour of strained atoms for example. Of course, if the dislocation emerges in a high index plane its core structure will, in principle, be visible. Table 5.1 summarises

TABLE 5.1

fcc lattice: values of $p$ calculated from eq. (5.1)

| hkl \ b | $\frac{1}{2}a[110]$ | $\frac{1}{2}a[101]$ | $\frac{1}{2}a[011]$ | $\frac{1}{2}a[\bar{1}10]$ | $\frac{1}{2}a[\bar{1}01]$ | $\frac{1}{2}a[0\bar{1}1]$ |
|---|---|---|---|---|---|---|
| (111) | 1 | 1 | 1 | 0 | 0 | 0 |
| ($\bar{1}$11) | 0 | 0 | 1 | 1 | 1 | 0 |
| (1$\bar{1}$1) | 0 | 1 | 0 | $-1$ | 0 | 1 |
| (11$\bar{1}$) | 1 | 0 | 0 | 0 | $-1$ | $-1$ |
| (200) | 1 | 1 | 0 | $-1$ | $-1$ | 0 |
| (020) | 1 | 0 | 1 | 1 | 0 | $-1$ |
| (002) | 0 | 1 | 1 | 0 | 1 | 1 |
| (220) | 2 | 1 | 1 | 0 | $-1$ | $-1$ |
| (202) | 1 | 2 | 1 | $-1$ | 0 | 1 |
| (022) | 1 | 1 | 2 | 1 | 1 | 0 |
| (2$\bar{2}$0) | 0 | 1 | $-1$ | $-2$ | $-1$ | 1 |
| (20$\bar{2}$) | 1 | 0 | $-1$ | $-1$ | $-2$ | $-1$ |
| (0$\bar{2}$2) | $-1$ | 1 | 0 | $-1$ | 1 | 2 |
| (113) | 1 | 2 | 2 | 0 | 1 | 1 |
| (131) | 2 | 1 | 2 | 1 | 0 | $-1$ |
| (311) | 2 | 2 | 1 | $-1$ | $-1$ | 0 |
| ($\bar{1}$13) | 0 | 1 | 2 | 1 | 2 | 1 |
| (13$\bar{1}$) | 2 | 0 | 1 | 1 | $-1$ | $-2$ |
| (3$\bar{1}$1) | 1 | 2 | 0 | $-2$ | $-1$ | 1 |
| (1$\bar{1}$3) | 0 | 2 | 1 | $-1$ | 1 | 2 |
| ($\bar{1}$31) | 1 | 0 | 2 | 2 | 1 | $-1$ |
| (31$\bar{1}$) | 2 | 1 | 0 | $-1$ | $-2$ | $-1$ |
| (11$\bar{3}$) | 1 | $-1$ | $-1$ | 0 | $-2$ | $-2$ |
| (1$\bar{3}$1) | $-1$ | 1 | $-1$ | $-2$ | 0 | 2 |
| ($\bar{3}$11) | $-1$ | $-1$ | 1 | 2 | 2 | 0 |
| (331) | 3 | 2 | 2 | 0 | $-1$ | $-1$ |
| (313) | 2 | 3 | 2 | $-1$ | 0 | 1 |
| (133) | 2 | 2 | 3 | 1 | 1 | 0 |
| (33$\bar{1}$) | 3 | 1 | 1 | 0 | $-2$ | $-2$ |
| (3$\bar{1}$3) | 1 | 3 | 1 | $-2$ | 0 | 2 |
| ($\bar{1}$33) | 1 | 1 | 3 | 2 | 2 | 0 |
| ($\bar{3}$31) | 0 | $-1$ | 2 | 3 | 2 | $-1$ |
| (31$\bar{3}$) | 2 | 0 | $-1$ | $-1$ | $-3$ | $-2$ |
| (1$\bar{3}$3) | $-1$ | 2 | 0 | $-2$ | 1 | 3 |
| (3$\bar{3}$1) | 0 | 2 | $-1$ | $-3$ | $-1$ | 2 |
| ($\bar{3}$13) | $-1$ | 0 | 2 | 2 | 3 | 1 |
| (13$\bar{3}$) | 2 | $-1$ | 0 | 1 | $-2$ | $-3$ |

Underlining indicates that an unambiguous assignment of Burgers vector is possible from a single micrograph.

the contrast expected from characteristic dislocations in the fcc lattice emerging in low index poles. Note that it is unusual to be able to identify the Burgers vector from any particular micrograph.

### 5.2.2. A SIGN CONVENTION

A positive line sense $l$ is arbitrarily chosen. Whilst looking along $l$ a clockwise closed circuit is performed in good crystal and repeated around the dislocation; for partial dislocations the circuit begins and ends on the stacking fault. The circuit fails to close and the vector $b$ joining the finish of the circuit to the start of the circuit is the Burgers vector of the dislocation. The vector $m = b \times l$ points to the extra half plane associated with the dislocation. Clearly if the opposite line sense $-l$ is chosen, the Burgers vector will be $-b$ and $m = b \times l$ will have the same sense. This property also enables a distinction to be made between intrinsic and extrinsic faults when $l$ and $b$ are known, (see § 5.4.1).

In field-ion images the line of a dislocation can be estimated from a field evaporation sequence (§ 1.8.4) or other crystallographic knowledge e.g. the plane of a faulted loop. The sense of the dislocation spiral indicates whether the component of $b$ in $n$ is parallel or anti-parallel. It is a convenient convention to take the positive sense of $l$ as that which has a component parallel to $n$; then a clockwise spiral, looking down on the micrograph, indicates $b$ resolved in $n$ to be antiparallel to $n$ i.e. $p$ or $q$ negative. An anti-clockwise spiral, looking down on the micrograph, indicates $p$ or $q$ to be positive. Raghaven (1967) has used a similar approach to determine the position of the extra half plane of an edge dislocation.

### 5.3. Contrast from perfect dislocation arrays

### 5.3.1. LOOPS

The quantity $p$ defined in eq. (5.1) can take both positive and negative integral values; in particular, dislocations having equal and opposite Burgers vectors and emerging in the same pole of a field-ion tip have $p$ values of equal magnitude but opposite sign. This means that the field-ion microscopic contrast from a dipole pair (equivalent to a loop intersecting the tip surface) can be understood as the sum of the opposite sense spiral effects associated with the two dislocations.

The simplest situation is when both dislocations emerge in the same pole. There are two cases to consider:
   a) the two dislocations end on the same imaged rings of atoms,
   b) the two dislocations end on separate imaged ring of atoms.

The resulting distortions to the specimen surface are sketched in figs. 5.3a and 5.3b respectively for $p = 1$. The spirals begun by the dislocations at A and C are ended by the dislocations emerging at B and D respectively. Any ring of atoms enclosing both dislocations is undisturbed, indicating

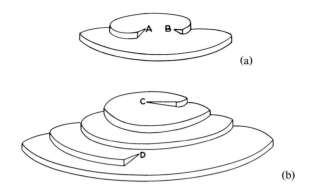

(a)

(b)

Fig. 5.3. Schematic diagrams of one pole of a field-ion tip intersected by a dislocation dipole such that: (a) both dislocations emerge in the same ledge, (b) the dislocations emerge in different ledges.

that the net Burgers vector of the dipole is zero. When $p > 1$ the contrast expected is similar combinations of multiple spirals. If the loop diameter is so great that the dipole dislocations emerge in separate poles then the contrast will be as that of two isolated dislocation lines. A clue suggesting that the dislocations are associated would be obtained from the sense of the spirals derived from the putative dipole pair, and the simultaneous appearance and disappearance of the two dislocations. Note that the spirals resulting from intersection of the tip surface by a dipole pair need not be of opposite sign, e.g. dislocations of Burgers vector $\pm\frac{1}{2}a[110]$ emerging in (202) and ($\bar{2}$02) respectively; $p = 1$ in both cases. If only geometrical effects are considered a clue as to the plane of the loops can be obtained by direct measurement of the line of intersection of the loop plane with the pole in which the dislocation emerges (§ 1.8.5).

### 5.3.2. DISLOCATION NETWORKS

It is very unlikely that field-ion microscopy will yield useful information regarding the bulk configuration of dislocation networks and tangles, because of the unknown influence of the field stress and the specimen surface. However, field-ion observations of some arrays of dislocations have been attributed to networks (Ranganathan, 1966b) and it is relevant to describe an example of the contrast to be expected in a field evaporation sequence through a triple node. Figure 5.4 is a diagram of a section through a field-ion tip which contains a dislocation network. The dashed lines indicate successive positions of the field evaporated surface. Initially a single

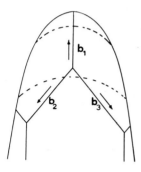

Fig. 5.4. Sketch of a section through a field-ion tip containing a dislocation node. The dashed lines indicate successive positions of the field-evaporated surface.

dislocation emerges in the imaged region of the field-ion tip whilst after field evaporation to the final dashed line two dislocations are released from what was a triple node. They may or may not take up positions in the imaged region of the tip surface. If they do, the simultaneous appearance of two dislocations immediately preceded by the disappearance of a third is strong circumstantial evidence that the dislocations belonged to a network. A check on this hypothesis is possible and entails testing the self consistency of the multiplicity and sense of the dislocation spirals, remembering that $b_1 + b_2 + b_3 = 0$ and are coplanar, if the network was formed by glide. If the dislocations continue to emerge in the same pole, the *net p* value is unchanged.

## 5.4. Contrast from partial dislocations and faults

### 5.4.1. STACKING FAULTS

The ideas of § 5.2 can also be applied to faults and partial dislocations. Consider a planar fault. The displacement at the fault is $R = u_p a + v_p b + w_p c$ where $u_p$, $v_p$ and $w_p$ are not all integers. The value of $R$ is the same everywhere on the surface of the fault. Let the fault lie on a plane of unit normal $s$; the case in which the plane of the fault lies in the imaged surface, of unit normal $n$ (i.e., where $n \times s = 0$), is excluded from the discussion which follows. In all other cases a step must be formed at the line of intersection of the fault and the imaged surface. This step is of height $R \cdot n$, normal to the imaged surface,

$$R \cdot n = q d_{hkl} \qquad (5.2)$$

where

$$q = h u_p + k v_p + l w_p; \qquad (5.3)$$

*q is not necessarily integral* (cf. *p* which must be integral). When *q* is greater than unity field evaporation is expected to remove an integral number of planes from one side of the step. When *q* is an integer no step remains after field evaporation. However contrast can still result from the fault; it is possible for the resolved component of the fault vector $R$ to produce a displacement in the rows of atoms making up the imaged plane edges. This form of contrast will be called a kink. It will be appreciated that a fault intersects a pole in a series of lines (each parallel to $n \times s$) during the course of field evaporation. When $q = 0$, $R$ is a vector in the plane of unit normal $n$. The displacement produced in a row of atoms in the direction of the unit vector $t$ in the plane of unit normal $n$ is of the magnitude $|R \times t|$ normal to $t$. Another case is when $q \neq 0$ but is an integer; then the fault can be described always by a vector $R_1$ differing from $R$ by an integral number of lattice vectors. The magnitude of the displacement normal to the rows is given by $|R_1 \times t|$. Clearly a kink will result only if $|R_1 \times t|$ or $|R \times t|$ does not equal an integral number of inter-row spacings. The magnitude of the kink produced normal to the rows is a quantity smaller than the inter-row separation, and is obtained by subtracting an integral number of inter-row spacings from $|R_1 \times t|$ or $|R \times t|$. Although the magnitude of the displacement is small, evidence is given in § 5.5.4 which shows that it is observable in a particular case in the field-ion image.

If *q* is not integral the step height after field evaporation is equal to the nonintegral residue, $q'$ (see fig. 5.5). On either side of the fault the unequal radii of the imaged "part planes" are expected (figs. 5.5c, 5.5d and 5.5e) because the imaged plane must conform to an approximately hemispherical envelope (fig. 5.1a). Continued field evaporation of the tip, at this stage, *cannot* remove the step. The contrast expected in the field-ion microscope is sketched in fig. 5.5e. Each plane on one side of the fault has two adjacent planes on the other side at distances along the local tip normal of $q' d_{hkl}$ and $(1 - q') d_{hkl}$. The distance of the topmost plane from the next (on the other side of the fault) will have one of these two values, and the particular value depends on the side of the fault on which the top-most plane lies, i.e.

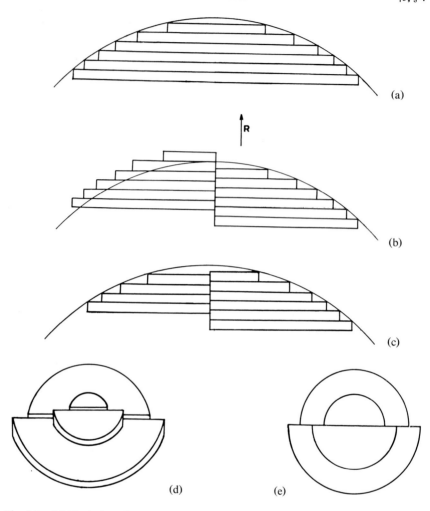

Fig. 5.5. (a) Vertical section through one pole of a field-ion tip when perfect. (b) As (a) after introduction of a stacking fault with $1 < |q| < 2$. (c) As (b) but after field evaporation leaving a step of height $|q'|$. (d) Sketch of surface, shown in section in (c). (e) Schematic diagram of contrast expected in the field-ion microscope from the surface shown in (d).

the distance will vary with field evaporation. Note that, in principle, it would be possible to distinguish two faults on parallel planes and intersecting the same pole but having opposite displacement vectors ($\boldsymbol{R}$ and $-\boldsymbol{R}$) by comparing ledge widths, and thus to make a distinction between an intrinsic and an extrinsic fault.

An example follows. Consider a fault in an fcc metal with $R = \frac{1}{6}a[112]$ lying in $(\bar{1}\bar{1}1)$ and emerging in $(013)$, then eq. 5.3 gives:

$$0 \cdot \tfrac{1}{6} + 2 \cdot \tfrac{1}{6} + 6 \cdot \tfrac{2}{6} = \tfrac{7}{3}.$$

Step heights of $\frac{1}{3}d_{026}$ and $\frac{2}{3}d_{026}$ are expected after field evaporation. The same value of $q'$ would be obtained *whatever* the value of $R$ used to define this particular fault.

A clue as to the plane on which the fault lies can be obtained from the direction of its trace in the imaged pole; in the example this would be $[\bar{4}3\bar{1}]$.

### 5.4.2. INTRINSIC AND EXTRINSIC FAULTS

The sign convention of § 5.2.2 provides one way of distinguishing intrinsic from extrinsic faults. However, the field-ion contrast may show differences which could be perceived immediately. The step height (i.e. the non-integral residue of $q$) due to a fault on a given plane intersecting a particular pole is the same whether the fault is intrinsic or extrinsic. This is illustrated in fig. 5.6 for a fault on $(11\bar{1})$ emerging in the $(002)$ pole of the image, which shows the atom arrangement in $(1\bar{1}0)$. Figure 5.6a is the perfect lattice; figs. 5.6b and 5.6c are the arrangements for intrinsic and extrinsic faults respectively. The difference in atomic arrangement at both steps and kinks could in principle lead to different contrast in the field-ion image. For example, if atoms 1, 2, 3 in fig. 5.6c were removed by field evaporation, a configuration at the step is produced which could not be obtained by evaporation of atoms in fig. 5.6b. The step heights can be the same in the two cases depending on the stage which field evaporation has reached i.e. $\frac{1}{3}$ and $\frac{2}{3}d_{002}$. It is possible, too, that vacancy and interstitial loops may be distinguished because of the opposite signs of their strain fields which could be detected as a bending of the imaged rings of atoms. There are no experimental data to decide these points. Computer simulations by Ranganathan (1969) consistent with the above ideas, predict that the field-ion image of an extrinsic fault in fcc materials would show two rows of atoms along the step at the fault in contrast to the single row predicted in the image of an intrinsic fault i.e. atoms $\alpha$ and $\beta$ in fig. 5.6c, but only atom $\gamma$ in fig. 5.6b are in the step associated with the stacking fault. Such fine detail may not be resolvable in practice.

### 5.4.3. PARTIAL DISLOCATIONS

A partial dislocation bounds a stacking fault. Both the effects due to a fault and to a dislocation will be observed at the partial dislocation. A spiral

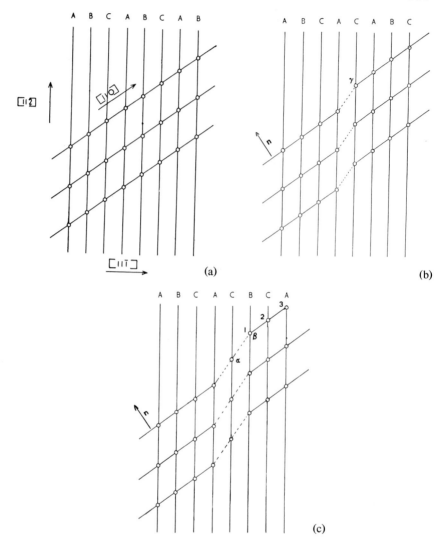

Fig. 5.6. (a) Plan of atom positions in the (1Ī0) plane of an fcc metal. (b) Plan of atom
positions in the (1Ī0) plane of an fcc metal containing an intrinsic fault on a (11Ī) plane.
(c) Plan of atom positions in the (1Ī0) plane of an fcc metal containing an extrinsic
fault on the (11Ī) plane.

will be produced of pitch equal to $n \cdot b_p$ where $b_p$ is the Burgers vector of the
partial dislocation. In terms of the spacing of the $(hkl)$ planes of unit normal
$n$ the pitch $q$ is given by eq. (5.2) where $R = b_p$. Consider first the case where

$0<|q|<1$ so that the pitch is less than an interplanar spacing. The resulting form of the surface is sketched in fig. 5.7. Note that successive turns of the spiral do not link up to form a continuous spiral. The resulting configuration is called a *stepped spiral*. It can happen that $q$ is a non-zero integer just as for

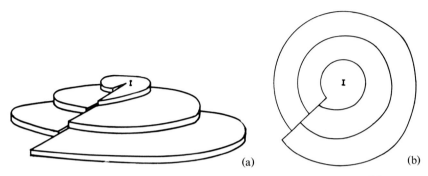

Fig. 5.7.  Schematic diagram of (a) the surface and (b) the contrast expected from one pole of a field-ion tip intersected by a partial dislocation emerging at I, with $0<|q|<1$ i.e. $q''=1$.

a total dislocation. In this case $q$ interleaved spirals are the result and each turn will be joined to the next along the line of intersection of the fault plane and the imaged pole by a kink due to displacements in the imaged pole as discussed in § 5.3. Such configurations are called *kinked spirals*. The kink permits a distinction to be made between the spirals from perfect and partial dislocations. Further when $q=0$ no spiral is seen but the displacements parallel to the imaged pole may make the fault visible by leading to kinked rings.

Another case is when $|q|>1$ and non-integral, when the tip surface becomes $q''$ interleaved stepped spirals, where $q''$ is the smallest integer greater than $|q|$. Figure 5.8 is a sketch of the surface and contrast expected in the field-ion image when $q''=2$. Note that, as for the perfect dislocation, the separation of the turns of the spiral is a consequence of the imaging criterion *not* of the small displacements parallel to the imaged pole.

The surface configuration resulting from intersection of a field-ion pole by faulted loops and extended total dislocations can now be deduced in terms of the preceding discussion. The crucial point is that the ring configuration seen in a field-ion micrograph e.g. spiral, stepped spiral, kinked rings, etc, is characteristic of the Burgers vector sum of the one or more dislocations enclosed by the imaged rings of atoms, and the displacements associated with the fault. Those rings which cross the fault are kinked or stepped. A dislocation loop intersecting the surface of a field-ion tip will give contrast equivalent to that from a dislocation dipole. The two dislocations which constitute a dipole have Burgers vectors equal in magnitude but opposite in sign. Hence when both dislocations are enclosed the imaged rings of atoms show no closure failure. The step configuration between the two dislocations depends on the value of $q$. Figure 5.9 is a sketch of the surface configuration for a faulted loop with $0 < |q| < 1$.

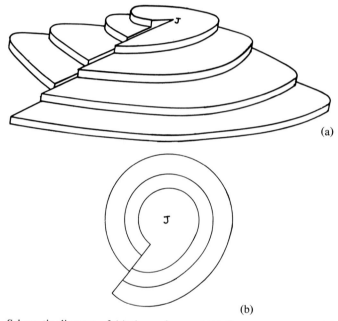

(a)

(b)

Fig. 5.8.   Schematic diagram of (a) the surface and (b) the contrast expected from one pole of a field-ion tip intersected by a partial dislocation emerging at J
with $1 < |q| < 2$ i.e. $q'' = 2$.

In principle there are many possible contrasts from dissociated total dislocations but in all cases these can be analyzed in terms of $p$ for the parent total dislocation and $q$ for the partial dislocations produced by dissocation. A particular example is shown in fig. 5.10 where $p = 1$ and $q = \frac{1}{3}$ and $\frac{2}{3}$ respectively for the partial dislocations emerging at H and G. When both partial dislocations are enclosed by an imaged ring of atoms the contrast is a simple spiral since $p = 1$. When only the partial dislocation emerging at H is enclosed $q = \frac{1}{3}$ and a single stepped spiral is seen. The partial dislocation emerging at G has $q = \frac{2}{3}$ which gives a *net* pitch of the spiral of one

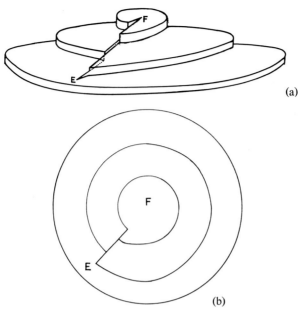

(a)

(b)

Fig. 5.9.   Schematic diagram of (a) the surface and (b) the contrast expected from one pole of a field-ion tip intersected by a faulted loop with $0 < |q| < 1$. The boundary partial dislocations emerge at E and F.

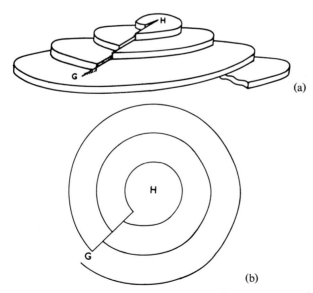

(a)

(b)

Fig. 5.10.   Schematic diagram of (a) the surface and (b) the contrast expected from one pole of a field-ion tip intersected by a dissociated total dislocation with $|p| = 1$.

interplanar spacing. If the pole illustrated in fig. 5.10 were (111) in an fcc metal, for example, then the dislocation reaction could be as follows:

$$\tfrac{1}{2}a[101]_{(11\bar{1})} \rightarrow \tfrac{1}{6}a[2\bar{1}1]_{(11\bar{1})} + \tfrac{1}{6}a[112]_{(11\bar{1})}$$
$$p = 1 \qquad\qquad q = \tfrac{1}{3} \qquad\qquad q = \tfrac{2}{3}.$$

Table 5.2 summarises typical contrasts to be expected from partial dislocations emerging in some prominent poles for fcc metals.

Table 5.3 indicates some situations where Frank and Shockley partial dislocations could be distinguished from a single observation.

TABLE 5.2

Typical contrast to be expected from partial dislocations emerging in some prominent poles for fcc metals

| Burgers vector | Pole | $q$ | Contrast |
|---|---|---|---|
| $\tfrac{1}{6}a[112]$ | (100) | $\tfrac{1}{3}$ | Stepped spiral |
| $\tfrac{1}{6}a[112]$ | (110) | $\tfrac{2}{3}$ | Stepped spiral |
| $\tfrac{1}{6}a[112]$ | (1$\bar{1}$0) | 0 | Kinked rings* |
| $\tfrac{1}{6}a[112]$ | (101) | 1 | Kinked spiral* |
| $\tfrac{1}{6}a[112]$ | (111) | $\tfrac{2}{3}$ | Stepped spiral |
| $\tfrac{1}{6}a[112]$ | (113) | $\tfrac{4}{3}$ | Double stepped spiral |
| $\tfrac{1}{3}a[111]$ | (100) | $\tfrac{2}{3}$ | Stepped spiral |
| $\tfrac{1}{3}a[111]$ | (110) | $\tfrac{4}{3}$ | Double stepped spiral |
| $\tfrac{1}{3}a[111]$ | (1$\bar{1}$0) | 0 | Kinked rings* |
| $\tfrac{1}{3}a[\bar{1}11]$ | (111) | $\tfrac{1}{3}$ | Stepped spiral |
| $\tfrac{1}{3}a[111]$ | (113) | $\tfrac{5}{3}$ | Double stepped spiral |
| $\tfrac{1}{3}a[\bar{1}11]$ | (113) | 1 | Kinked spiral* |

* The kink part of the contrast depends on which of the atomic rows within the imaged pole are intersected by the fault associated with these dislocations.

TABLE 5.3

$q$ values for Frank and Shockley partials; unambiguous cases are underlined

| | $\tfrac{1}{3}a[111]$ | $\tfrac{1}{6}a[\bar{1}\bar{1}2]$ | $\tfrac{1}{6}a[\bar{1}2\bar{1}]$ | $\tfrac{1}{6}a[2\bar{1}\bar{1}]$ |
|---|---|---|---|---|
| (hkl) | | | | |
| 113 | $\tfrac{5}{3}$ | $\tfrac{2}{3}$ | $-\tfrac{1}{3}$ | $-\tfrac{1}{3}$ |
| 131 | $\tfrac{5}{3}$ | $-\tfrac{1}{3}$ | $\tfrac{2}{3}$ | $-\tfrac{1}{3}$ |
| 311 | $\tfrac{5}{3}$ | $-\tfrac{1}{3}$ | $-\tfrac{1}{3}$ | $\tfrac{2}{3}$ |
| 11$\bar{3}$ | $-\tfrac{1}{3}$ | $-\tfrac{4}{3}$ | $\tfrac{2}{3}$ | $\tfrac{2}{3}$ |
| 1$\bar{3}$1 | $-\tfrac{1}{3}$ | $\tfrac{2}{3}$ | $-\tfrac{4}{3}$ | $\tfrac{2}{3}$ |
| $\bar{3}$11 | $-\tfrac{1}{3}$ | $\tfrac{2}{3}$ | $\tfrac{2}{3}$ | $-\tfrac{4}{3}$ |
| $\bar{1}$13 | 1 | 1 | 0 | $-\bar{1}$ |
| 13$\bar{1}$ | 1 | $-1$ | 1 | 0 |
| 3$\bar{1}$1 | 1 | 0 | $-1$ | 1 |
| 1$\bar{1}$3 | 1 | 1 | $-1$ | 0 |
| $\bar{1}$31 | 1 | 0 | 1 | $-\bar{1}$ |
| 31$\bar{1}$ | 1 | $-1$ | 0 | 1 |

## 5.5. Dislocation observations

### 5.5.1. SINGLE PERFECT DISLOCATIONS

Figure 5.11 shows a single perfect dislocation emerging in the $\{111\}$ pole of an iridium specimen. Since in cubic crystals all $\{hkl\}$ planes are equivalent, we can call this particular pole (111). A single anticlockwise spiral is apparent, i.e. $p = +1$. The expected perfect dislocation Burgers vectors in the fcc lattice are $\frac{1}{2}a\langle110\rangle$ i.e. the possible vectors are $\pm\frac{1}{2}a[110]$,

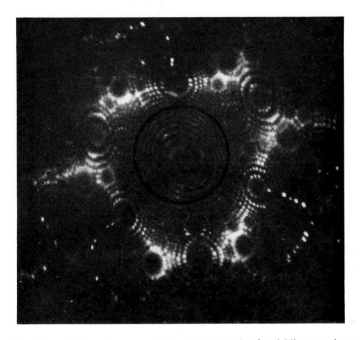

Fig. 5.11.  A dislocation emerging in the (111) pole of an iridium specimen: 78°K, helium image gas.

$\pm\frac{1}{2}a[011]$, $\pm\frac{1}{2}a[101]$, $\pm\frac{1}{2}a[1\bar{1}0]$, $\pm\frac{1}{2}a[01\bar{1}]$, and $\pm\frac{1}{2}a[\bar{1}01]$. The last six possibilities can be excluded since $p = 0$ in each case. However, each of the first six possibilities gives $p = \pm 1$, and further characterisation is not possible since this particular dislocation moved out of the tip after a small field evaporation, apart from rejection of the vectors giving $p = -1$ because of the sense of the spiral (see § 5.2.2; sign convention).

Figure 5.12 shows a two-leaved anticlockwise spiral emerging in the

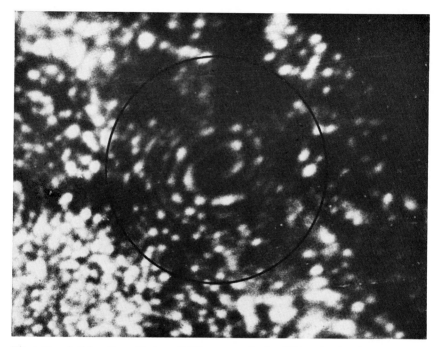

Fig. 5.12. A dislocation with $b = a[110]$ emerging in the (110) pole of an iron specimen. Note the two leaved spiral (78 °K, hydrogen image gas). (Courtesy R. Morgan.)

(110) pole of an iron field-ion specimen. The only single dislocation Burgers vector which generates this contrast is $a[110]$.

Figure 5.13 shows another perfect dislocation emerging in the (331) pole of an iridium field-ion specimen. The triple clockwise spiral shows that $p = -3$. The only possible Burgers vector of the $\frac{1}{2}a\langle 110 \rangle$ type for which $p = -3$ is $\frac{1}{2}a[\overline{1}\overline{1}0]$ i.e. in this case the Burgers vector can be determined from a single observation.

### 5.5.2. PERFECT DISLOCATION LOOPS

Figures 5.14a and 5.14b illustrate perfect dislocation loops emerging in the (113) pole of an iridium field-ion specimen, in the orientations of figs. 5.3a and 5.3b. Perfect dislocation loops in fcc metals such as iridium are expected to occur on {110} or {111} type planes. In fig. 5.14a, the line of intersection of the loop plane is $[33\overline{2}]$ corresponding to a loop on $(1\overline{1}0)$. In fig. 5.14b the line of intersection of the loop plane is perpendicular to $[33\overline{2}]$ i.e. $[1\overline{1}0]$, corresponding to a loop on (110), (111), or $(11\overline{1})$. Taking the

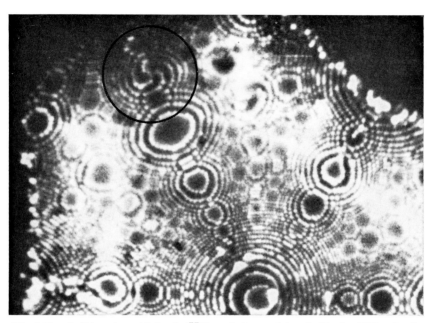

Fig. 5.13.   A dislocation with $b = \frac{1}{2}a[\bar{1}\bar{1}0]$ producing a triple spiral in the (331) pole of an iridium field-ion specimen (78°K, helium image gas). (Courtesy T. F. Page.)

ends of the spirals as the edges of the loops the apparent diameter of the loops in figs. 5.14a and 5.14b can be estimated by comparison with calculated ledge widths and known interatomic separations as about 27 Å and 36 Å respectively. The Burgers vectors of the loops cannot be determined unambiguously. Initially taking the positive values of the vectors the six possibilities are:

$$\tfrac{1}{2}a[110], \tfrac{1}{2}a[011], \tfrac{1}{2}a[101], \tfrac{1}{2}a[1\bar{1}0], \tfrac{1}{2}a[0\bar{1}1], \quad \text{and} \quad \tfrac{1}{2}a[\bar{1}01]$$

giving $p = 1, 2, 2, 0, 1, 1$ i.e. the Burgers vectors $\tfrac{1}{2}a[110]$, $\tfrac{1}{2}a[0\bar{1}1]$, and $\tfrac{1}{2}a[\bar{1}01]$ are equally consistent with the contrast observed in figs. 5.3a and 5.3b. As an example we apply the sign convention of § 5.2.2 to the dislocation emerging in the central ring of fig. 5.14b. If we take $l = [001]$ then $b = \tfrac{1}{2}a[\bar{1}\bar{1}0]$ and $b \times l = [\bar{1}10]$ i.e. the loop is interstitial in nature. The dimensions normal to the tip surface of a loop, and its shape can be determined from an atom by atom field evaporation sequence.

### 5.5.3. Dissociated perfect dislocations and faulted loops

Figures 5.15a, 5.15b and 5.15c are micrographs showing a perfect

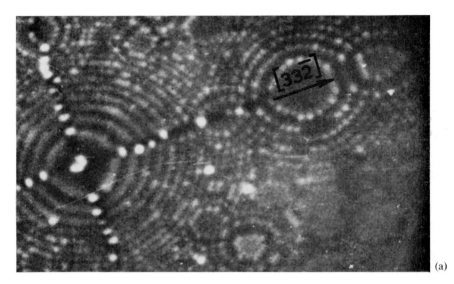

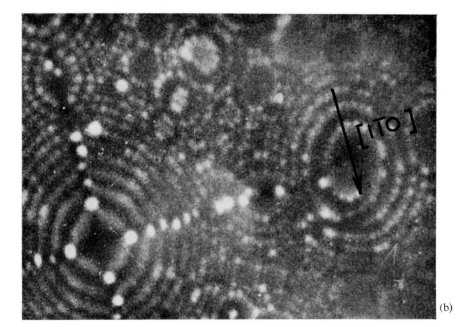

Fig. 5.14.  (a) and (b). Perfect dislocation loops emerging in the (113) poles of iridium
specimens, cf. 5.3 (a) and (b) (78°K, helium image gas). (Courtesy M. A. Fortes.)

dislocation emerging in the (111) pole of an iridium field-ion tip. These micrographs are part of an atom by atom field evaporation sequence. Less than one (111) plane has been field evaporated in this sequence. The configuration in figs. 5.15a and 5.15c is a clockwise simple spiral giving $p = -1$, whilst that in fig. 5.15b is initially a stepped spiral having $0 < |q| < 1$. The trace of the fault plane in (111) corresponds closely to $[\bar{1}01]$. A $(1\bar{1}1)$ fault plane is consistent with a $[\bar{1}01]$ trace. The three $\frac{1}{2}a\langle 110 \rangle$ type Burgers vectors giving $p = -1$ in (111) are:

$$\tfrac{1}{2}a[\bar{1}\bar{1}0], \quad \tfrac{1}{2}a[0\bar{1}\bar{1}], \quad \text{and} \quad \tfrac{1}{2}a[\bar{1}0\bar{1}].$$

A dislocation with a Burgers vector $\frac{1}{2}a[\bar{1}0\bar{1}]$ can, however, be excluded as a possible source of the contrast in fig. 5.15b since it cannot dissociate on the $(1\bar{1}1)$ plane because the vector does not lie on this plane. The following dissociations are equally probable:

$$\tfrac{1}{2}a[\bar{1}\bar{1}0] \rightarrow \tfrac{1}{6}a[\bar{1}2\bar{1}]_{(1\bar{1}1)} + \tfrac{1}{6}a[\bar{2}\bar{1}1]_{(1\bar{1}1)}$$
$$p = -1 \quad q = -\tfrac{2}{3} \quad q = -\tfrac{1}{3}$$

or

$$\tfrac{1}{2}a[0\bar{1}\bar{1}] \rightarrow \tfrac{1}{6}a[\bar{1}2\bar{1}]_{(1\bar{1}1)} + \tfrac{1}{6}a[1\bar{1}\bar{2}]_{(1\bar{1}1)}$$
$$p = -1 \quad q = -\tfrac{2}{3} \quad q = -\tfrac{1}{3}.$$

Figures 5.15a and 5.15c show (Smith et al., 1969) that the ribbon width, $w$, is less than the width of the ledge in which the dislocation emerges (i.e. less than 20 Å). A crude estimate of the stacking fault energy of iridium can be made from the ribbon width. Some allowance should be made for the influence of the surface on $w$, following Gilman's (1962) use of a relation derived by Eshelby and Stroh (1951) for screw dislocations dissociated into Shockley particles. The stacking fault energy, $\gamma$, is estimated to be $\approx 125$ ergs·cm$^{-2}$. The field stress could cause either extension or contraction of the ribbon.

Figures 5.16 and 5.17 show the two main types of contrast predicted for faulted dislocation loops, stepped spiral and kinked spiral respectively. In each case the apparent size of the defects increased and decreased as material was removed by field evaporation; finally they disappeared. There is no net spiral when the defect is enclosed by an imaged ring of atoms which excludes the possibility that the defects are dissociated dislocations with $p \neq 0$. Figure 5.16 shows an example of fault contrast in the (111) pole of an iridium specimen. Two turns of an anticlockwise stepped spiral are visible together with a line of contrast as expected where the stacking fault intersects the tip surface (see fig. 5.9) which was found by direct measurement to be $[0\bar{1}\bar{1}]$; hence the stacking fault plane is $(\bar{1}11)$. The possible Burgers vectors

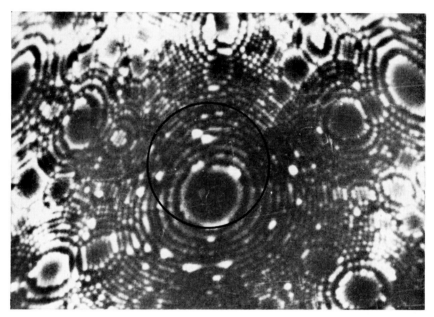

Fig. 5.15a

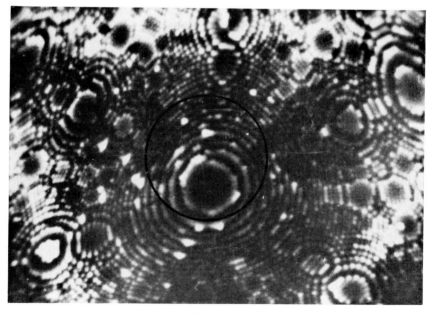

Fig. 5.15b

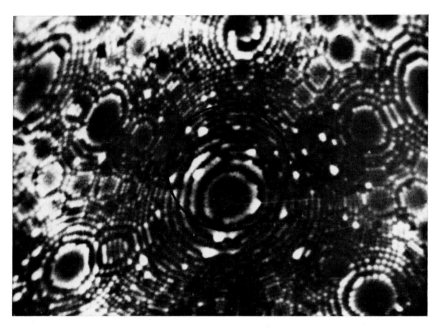

Fig. 5.15c

Fig. 5.15.   (a), (b) and (c). Successive micrographs recorded in an "atom by atom" field evaporation sequence of a specimen with a dislocation emerging in the (111) pole. The dislocation appears undissociated, dissociated and undissociated, successively (see text), (78°K, helium image gas). (Courtesy T. F. Page.)

of total dislocations in the $(\bar{1}11)$ plane are $\frac{1}{2}a[101]$, $\frac{1}{2}a[01\bar{1}]$ and $\frac{1}{2}a[110]$. Since there is no spiral in fig. 5.16 when both the dislocations bounding the stacking fault are enclosed $p=0$ for a putative dissociated total dislocation. This consideration excludes $\frac{1}{2}a[101]$ and $\frac{1}{2}a[110]$ as possible Burgers vectors. It is possible that the defect in fig. 5.16 is a total dislocation dissociated in $(\bar{1}11)$ as follows;

$$\frac{1}{2}a[01\bar{1}] \rightarrow \frac{1}{6}a[\bar{1}1\bar{2}]_{(\bar{1}11)} + \frac{1}{6}a[12\bar{1}]_{(\bar{1}11)}$$
$$p = 0 \qquad q = -\tfrac{1}{3} \qquad q = \tfrac{1}{3}.$$

Alternatively, since the fault plane is $(\bar{1}11)$ the Burgers vector of a Frank sessile dislocation loop would be $\frac{1}{3}a[\bar{1}11]$, which gives $q = +\tfrac{1}{3}$ and is consistent with the observed contrast.

It is possible that under the very high shear stresses present during imaging of a field-ion tip, Shockley loops may be nucleated. Such loops, with

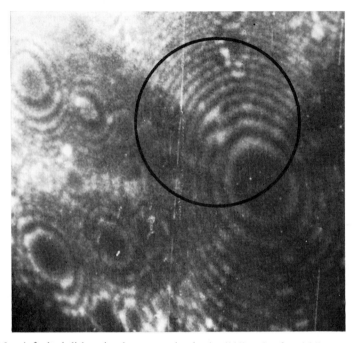

Fig. 5.16.   A faulted dislocation loop emerging in the (111) pole of an iridium specimen
(78 °K, helium image gas). (Courtesy M. A. Fortes.)

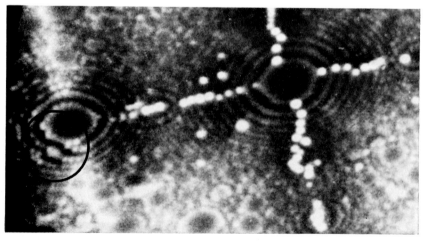

Fig. 5.17.   A faulted dislocation loop emerging in the (113) pole of an iridium specimen
(78 °K, helium image gas). (Courtesy M. A. Fortes.)

a Burgers vector $\frac{1}{6}a\langle112\rangle$, could grow and in special cases be stable for particular orientations. It is therefore relevant to consider the possibility that the defect in fig. 5.16 is a faulted loop bounded by Shockley partial dislocations. The three possible Burgers vectors of this type in the $(\bar{1}11)$ plane are $\frac{1}{6}a[211]$, $\frac{1}{6}a[12\bar{1}]$ and $\frac{1}{6}a[1\bar{1}2]$ which have $q$ values $+\frac{2}{3}$, $+\frac{1}{3}$, and $+\frac{1}{3}$ respectively in (111).

Each $q$ value leads to a stepped spiral type contrast and it is not possible in this case to distinguish between the three possibilities: dissociated total dislocation, Shockley loop or Frank loop. Note that the Frank and Shockley Burgers vectors do not differ by a lattice vector and therefore describe different faults the atomic differences of which cannot be discerned.

Figure 5.17 shows a partial dislocation arrangement in the (113) pole. The line of intersection of the fault is $[12\bar{1}]$, which means that the fault plane is $(\bar{1}11)$. Dissociated total dislocations with possible Burgers vectors $\frac{1}{2}a[110]$, $\frac{1}{2}a[101]$, and $\frac{1}{2}a[01\bar{1}]$ are rejected as potential sources of the contrast since $p \neq 0$ in each case and because the spiral in the figure stops after two turns anticlockwise i.e. the sum of the displacements in $\boldsymbol{n}$ due to the dislocations involved is zero. The kink where the imaged rings of atoms cross the $[12\bar{1}]$ direction indicates that the loop is faulted. Hence the Burgers vector of a Frank dislocation would be $\frac{1}{3}a[\bar{1}11]$ and $q = +1$, which gives a single kinked spiral. The possible Shockley bounding dislocations are $\frac{1}{6}a[2\bar{1}\bar{1}]$, $\frac{1}{6}a[12\bar{1}]$, and $\frac{1}{6}a[1\bar{1}2]$ which give $q$ values $-1$, 0 and $+1$ respectively, i.e. the observed single kinked spiral is consistent with a Frank loop or a Shockley loop with Burgers vector $\frac{1}{6}a[1\bar{1}2]$. The fault intersects the (113) pole where the planes are bounded by $[\bar{3}01]$ rows. The calculated magnitude of the kink is the same in each case and alternately takes the values $\frac{1}{3}\delta_{\bar{3}01}$ and $\frac{2}{3}\delta_{\bar{3}01}$ where $\delta_{\bar{3}01}$ is the spacing between the $[\bar{3}01]$ rows in the (113) plane of iridium (about 2 Å).

The result is calculated as follows. It is necessary to find an equivalent vector description of the fault such that $\boldsymbol{R}_1$ lies in (113). See § 5.4.1. The required vector is $\frac{1}{6}a[12\bar{1}]$ (note that $\frac{1}{6}a[12\bar{1}]$ differs from $\frac{1}{6}a[1\bar{1}2]$ by $\frac{1}{2}a[01\bar{1}]$). Now if $t$ is $10^{-\frac{1}{2}}[\bar{3}01]$

$$|\boldsymbol{R}_1 \times \boldsymbol{t}| = \left| \frac{a}{6\sqrt{10}}[226] \right| = \frac{1}{3}a\sqrt{\frac{11}{10}}.$$

Now $\delta_{\bar{3}01} = \frac{1}{2}a\sqrt{11/10}$. Therefore the non-integral residue of $|\boldsymbol{R}_1 \times \boldsymbol{t}|$ is $\frac{1}{3}\delta_{\bar{3}01}$.

Fortes and Ralph (1968a) have positively identified Shockley loops in the same sequence of field-ion micrographs from which figs. 5.16 and 5.17

are taken. These have been produced by homogeneous nucleation of dislocations in a fracture of the specimen at the theoretical stress (Frank, 1950; Kelly, 1966), attained under the influence of the applied field. Figure 5.18 shows a faulted loop emerging in the (131) pole of an iridium field-ion specimen. The fault plane is $(1\bar{1}1)$ in which the only Frank or Shockley vector giving $q'' = +2$ is $\frac{1}{6}a\,[121]$.

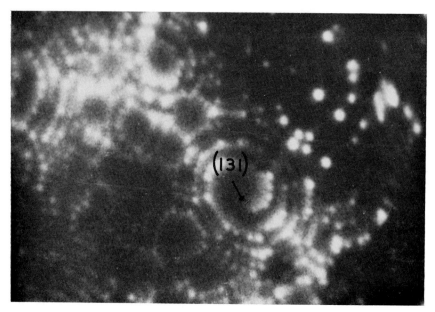

Fig. 5.18.   A Shockley loop emerging in the (131) pole of an iridium specimen (78°K, helium image gas). (Courtesy M. A. Fortes.)

5.5.4. EXTENDED STACKING FAULT

Figure 5.19 shows an example of a stacking fault intersecting the (013) pole of an iridium specimen cf. fig. 5.5e. The differences between geometrical prediction and observation can be accounted for in terms of secondary effects (§ 5.7). The line of intersection of the fault plane and (013) is $[23\bar{1}]$, corresponding to a fault on $(\bar{1}11)$. The residual step height caused by a fault on $(\bar{1}11)$ emerging in (013) is $\frac{1}{3}d_{(026)}$ or $\frac{2}{3}d_{(026)}$; eqs. (5.2) and (5.3). The lack of resolution precludes characterisation of the bounding partial dislocations.

## 5.6.  Stacking faults and partial dislocations in bcc materials

Partial dislocation configurations in bcc materials can be analysed following the same general method as that outlined in respect of fcc metals, i.e. determination of $p$ for the parent perfect dislocation, $q$ for partial dislocations and $f$ the line of intersection of the fault plane with the field-ion

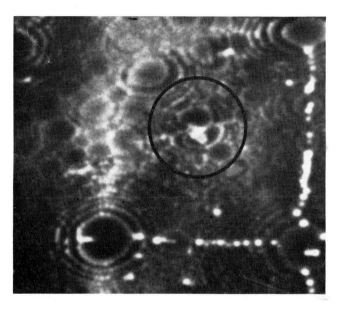

Fig. 5.19.   An extended stacking fault emerging in the (013) pole of an iridium specimen (78 °K, helium image gas). (Courtesy M. A. Fortes.)

tip. An exhaustive treatment of stacking faults in bcc is lengthy and complicated and we shall confine ourselves to a few examples illustrating new features of dislocation contrast. Full discussions are to be found in the papers of Smith and Bowkett (1968b) and Smith (1969b).

A number of dislocation dissociations have been suggested to account for the mechanical properties and phase transformations of bcc metals.

The following dislocation reaction in bcc metals which gives a stacking fault on (112) was proposed by Crussard (1961) and Sleeswyk and Verbraak (1961):

$$\tfrac{1}{2}a\left[11\bar{1}\right] \to \tfrac{1}{3}a\left[11\bar{1}\right]_{(112)} + \tfrac{1}{6}a\left[11\bar{1}\right]_{(112)}.$$

A screw dislocation can in principle dissociate on three different $\{112\}$ type planes as follows to give three twin faults:

$$\tfrac{1}{2}a\,[11\bar{1}] \to \tfrac{1}{6}a\,[11\bar{1}]_{(\bar{1}\bar{1}2)} + \tfrac{1}{6}a\,[11\bar{1}]_{(\bar{1}21)} + \tfrac{1}{6}a\,[11\bar{1}]_{(2\bar{1}1)}.$$

Sleeswyk (1963) showed that the lowest energy configuration is that where the dislocations on say $(\bar{1}\bar{1}2)$ and $(2\bar{1}1)$ have separated leaving two stacking fault ribbons with the third $\tfrac{1}{6}a\,[11\bar{1}]$ partial dislocation at the intersection of these planes. Dissociations on $\{110\}$ planes in bcc metals have also been suggested. Crussard (1961) and Cohen, Hinton, Lay and Sass (1962) have proposed a dissociation on the $(1\bar{1}0)$ plane:

$$\tfrac{1}{2}a\,[111] \to \tfrac{1}{8}a\,[110]_{(1\bar{1}0)} + \tfrac{1}{4}a\,[112]_{(1\bar{1}0)} + \tfrac{1}{8}a\,[110]_{(1\bar{1}0)}.$$

Kroupa (1963) and Kroupa and Vitek (1964) have suggested a further possible dissociation for screw dislocations to be:

$$\tfrac{1}{2}a\,[111] \to \tfrac{1}{8}a\,[101]_{(\bar{1}01)} + \tfrac{1}{8}a\,[110]_{(1\bar{1}0)} + \tfrac{1}{8}a\,[011]_{(01\bar{1})} + \tfrac{1}{4}a\,[111],$$

where the three $\tfrac{1}{8}a\langle110\rangle$ type dislocations separate on different $\{110\}$ type planes.

Wasilewski (1965) also proposed a dissociation on three $\{110\}$ type planes:

$$\tfrac{1}{2}a\,[111] \to \tfrac{1}{6}a\,[111]_{(1\bar{1}0)} + \tfrac{1}{6}a\,[111]_{(\bar{1}01)} + \tfrac{1}{6}a\,[111]_{(01\bar{1})}.$$

Foxall, Duesbery and Hirsch (1967) remark that this dissociation would occur with a greater reduction in energy on one $\{110\}$ type plane, and propose a number of reactions involving dissociation on $\{110\}$ and $\{112\}$, e.g.

$$\tfrac{1}{2}a\,[111] \to \tfrac{1}{6}a\,[111]_{(11\bar{2})} + \tfrac{1}{6}a\,[111] + \tfrac{1}{6}a\,[111]_{(1\bar{1}0)}.$$

The first $\tfrac{1}{6}a\,[111]$ partial lies on the $(11\bar{2})$ plane, the third partial lies on the $(1\bar{1}0)$ plane and the second partial lies at the intersection of $(1\bar{1}0)$ and $(11\bar{2})$ i.e. along the $[111]$ direction.

The field-ion contrast from dislocations dissociated according to the reactions discussed above or indeed other reactions can be described in terms of the parameters $p$ and $q$ and the line of intersection of the fault plane with the pole in which the defect emerges.

The surface configuration and expected field-ion contrast produced by a dislocation dissociated according to (a) are shown in fig. 5.10. If the indices $(110)$ are assigned to the pole in which the extended dislocation emerges then the reaction could be:

$$\tfrac{1}{2}a\,[111] \rightarrow \tfrac{1}{6}a\,[111]_{(11\bar{2})} + \tfrac{1}{3}a\,[111]_{(11\bar{2})}$$
$$p = 1 \qquad q = \tfrac{1}{3} \qquad q = \tfrac{2}{3}, \qquad\qquad\qquad (a)$$

the line of intersection of the fault plane with (110) is $[1\bar{1}0]$ and could be identified on a micrograph to demonstrate that the contrast is consistent with a $(11\bar{2})$ fault plane.

A typical surface configuration and field-ion contrast produced by a coplanar array of three dislocations is shown in fig. 5.20. The actual dislocation reactions, if the pole illustrated is (121), could be:

$$\tfrac{1}{2}a\,[11\bar{1}] \rightarrow \tfrac{1}{6}a\,[11\bar{1}]_{(112)} + \tfrac{1}{6}a\,[11\bar{1}]_{(112)} + \tfrac{1}{6}a\,[11\bar{1}]_{(112)}$$
$$p = 1 \qquad q = \tfrac{1}{3} \qquad q = \tfrac{1}{3} \qquad q = \tfrac{1}{3}, \qquad (b)$$

$$\tfrac{1}{2}a\,[11\bar{1}] \rightarrow \tfrac{1}{8}a\,[110]_{(1\bar{1}0)} + \tfrac{1}{4}a\,[11\bar{2}]_{(1\bar{1}0)} + \tfrac{1}{8}a\,[110]_{(1\bar{1}0)}$$
$$p = 1 \qquad q = \tfrac{3}{8} \qquad q = \tfrac{1}{4} \qquad q = \tfrac{3}{8}, \qquad (c)$$

$$\tfrac{1}{2}a\,[11\bar{1}] \rightarrow \tfrac{1}{6}a\,[11\bar{1}]_{(1\bar{1}0)} + \tfrac{1}{6}a\,[11\bar{1}]_{(1\bar{1}0)} + \tfrac{1}{6}a\,[11\bar{1}]_{(1\bar{1}0)}$$
$$p = 1 \qquad q = \tfrac{1}{3} \qquad q = \tfrac{1}{3} \qquad q = \tfrac{1}{3}. \qquad (d)$$

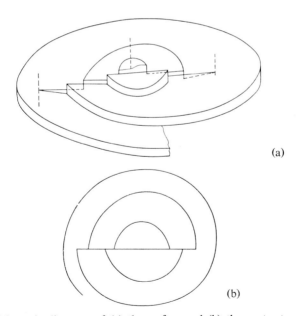

(a)

(b)

Fig. 5.20. Schematic diagrams of (a) the surface and (b) the contrast expected from one pole of a field-ion tip intersected by a dislocation dissociated according to e.g. the Cohen-Crussard reaction. An example of the contrast in a field-ion micrograph is shown in fig. 5.24.

Reaction (b) could generally be distinguished from reactions (c) and (d), by noting that $\{112\}$ and $\{110\}$ planes do not intersect the pole concerned in the same line. In the example given above $(1\bar{1}0)$ intersects $(121)$ in $[11\bar{3}]$ whilst $(112)$ intersects $(121)$ in $[\bar{3}11]$. $[\bar{3}11]$ and $[11\bar{3}]$ are inclined at $\approx 117°$. It is relevant to mention that the relative radii of the ledges of a field-ion pole on either side of a stacking fault depend on the stacking fault shear. If the non-integral residue of $q$ is $q'$, then the radii of the $i$-th ledges on either side of the fault are approximately:

$$\sqrt{2Rdi} \quad \text{and} \quad \sqrt{2Rd(i + q')}, \tag{5.4}$$

where $R$ is the tip radius and $d$ is the appropriate interplanar spacing. This is an extension of the earlier use of this relationship by Fortes and Ralph. The point of emergence of the middle dislocation in a coplanar array such as is considered above is not obvious. However, its presence can be inferred from the change in shape of the ledge enclosing it. This is required in order that material having been displaced along the normal to the pole concerned, the tip shape may still conform to a smoothly curved envelope. If, for example, the ledge was originally circular its radius would no longer be constant when a dislocation was enclosed. In practice it is not possible in the example given above to distinguish reactions (c) and (d) from a consideration of relative ledge widths. Use of (5.4) indicates that the relative radii of the top pair of part planes, assumed to cross the fault between the first pair of dislocations, are in the ratio

$$\frac{\sqrt{2Rd}}{\sqrt{2Rd\frac{4}{3}}} : \frac{\sqrt{2Rd}}{\sqrt{2Rd\frac{11}{8}}} \approx 1.01:1,$$

i.e. the expected difference in ledge width is only 1% which might typically be 1 Å and irresolvable by field-ion microscopy.

The presence and approximate location of the middle dislocation are quite obvious in principle when it marks the boundary between stacking faults which are not coplanar. Figure 5.21 shows a) the surface and b) the contrast expected from one pole of a field-ion tip intersected by a dislocation dissociated according to Sleeswyk's dissociation, equation (e). The actual reaction in the (121) pole, for example, could be:

$$\tfrac{1}{2}a[11\bar{1}] \to \tfrac{1}{6}a[11\bar{1}]_{(112)} + \tfrac{1}{6}a[11\bar{1}]_{[11\bar{1}]} + \tfrac{1}{6}a[11\bar{1}]_{(2\bar{1}1)}$$
$$p = 1 \qquad q = \tfrac{1}{3} \qquad q = \tfrac{1}{3} \qquad q = \tfrac{1}{3}, \tag{e}$$

Note the inclined steps on (112) and $(2\bar{1}1)$ which intersect (121) in $[\bar{3}11]$ and $[31\bar{5}]$ respectively. The contrast from other reactions involving two fault planes would be similar e.g., a reaction of type (f) where the partial dislocations might emerge in (110) giving the following $p$ and $q$ values consistent with fig. 5.21:

$$\tfrac{1}{2}a\,[111] \rightarrow \tfrac{1}{6}a\,[111]_{(11\bar{2})} + \tfrac{1}{6}a\,[111]_{[111]} + \tfrac{1}{6}a\,[111]_{(1\bar{1}0)}$$
$$p = 1 \qquad q = \tfrac{1}{3} \qquad\qquad q = \tfrac{1}{3} \qquad\qquad q = \tfrac{1}{3}. \tag{f}$$

In this example the lines of intersection of the $(11\bar{2})$ and $(1\bar{1}0)$ fault planes could be directly determined from a micrograph as $[1\bar{1}0]$ and $[001]$ respectively. In general it would be possible to distinguish dislocations dissociated according to reaction (e) and reaction (f) by noting the lines of intersection of the fault planes which are both $\{112\}$ in the first instance and $\{110\}$ and $\{112\}$ in the second. A distinction between reactions giving composite

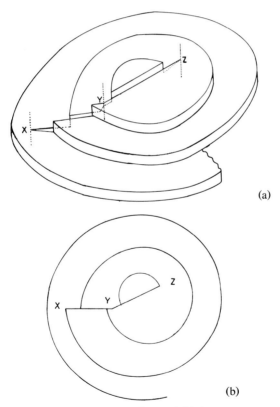

(a)

(b)

Fig. 5.21.   Schematic diagrams of (a) the surface and (b) the contrast expected from one pole of a field-ion tip intersected by a composite dislocation. An example of this contrast in a field-ion micrograph is shown in fig. 5.25.

dislocations of different Burgers vector could be attempted by calculating ledge widths using eq. (5.4).

Figure 5.22 shows a) the surface and b) the expected field-ion contrast from one pole of a field-ion tip intersected by a dislocation dissociated according to reactions of type (g). If the pole shown in fig. 5.22 is (110) a reaction giving the contrast is as follows:

$$\tfrac{1}{2}a\,[111] \rightarrow \tfrac{1}{6}a\,[111]_{(11\bar{2})} + \tfrac{1}{6}a\,[111]_{(1\bar{2}1)} + \tfrac{1}{6}a\,[111]_{(\bar{2}11)}$$

$$p = 1 \qquad q = \tfrac{1}{3} \qquad\qquad q = \tfrac{1}{3} \qquad\qquad q = \tfrac{1}{3}, \tag{g}$$

where the three inclined $(11\bar{2})$, $(1\bar{2}1)$ and $(\bar{2}11)$ faults planes would be visible with $[1\bar{1}0]$, $[\bar{1}13]$, and $[\bar{1}1\bar{3}]$ lines of intersection with (110).

The contrast expected from a dislocation dissociated according to (h)

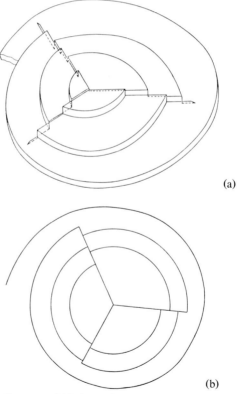

(a)

(b)

Fig. 5.22.   Schematic diagrams of (a) the surface and (b) the contrast expected from one pole of a field-ion tip intersected by a dislocation dissociated to give three non-coplanar faults.

is similar to that shown in fig. 5.22 but the ledge width increases will be modified owing to the different $q$ values, e.g. in (110),

$$\tfrac{1}{2}a[111] \rightarrow \tfrac{1}{8}a[101]_{(\bar{1}01)} + \tfrac{1}{8}a[011]_{(01\bar{1})} + \tfrac{1}{8}a[110]_{(1\bar{1}0)} + \tfrac{1}{4}a[111]$$

$$p = 1 \qquad q = \tfrac{1}{8} \qquad q = \tfrac{1}{8} \qquad q = \tfrac{1}{4} \qquad q = \tfrac{1}{2} \qquad \text{(h)}$$

and the faults are on {110} planes. The dislocation at the intersection of the fault planes does not generate a step and would not therefore give any obvious field-ion contrast unless the atomic arrangement was resolved. Geometrically it serves to make the steps generated by the $\tfrac{1}{8}a\langle 110\rangle$ dislocations into a whole number of interplanar spacings.

When $p = 0$ no spiral is expected, though selective field evaporation may nevertheless make the dislocation visible. When a dissociated dislocation with $p = 0$ emerges in a field-ion pole contrast is expected between the partial dislocations where an imaged ring of atoms crosses the fault plane. Figure 5.23 is an example showing the surface and expected contrast where a dislocation dissociated according to (c) emerges in one pole of a field-ion tip. If the pole concerned is taken to be (121) then the dislocation reaction could be as follows:

$$\tfrac{1}{2}a[1\bar{1}1] \rightarrow \tfrac{1}{8}a[1\bar{1}0] + \tfrac{1}{4}a[1\bar{1}2] + \tfrac{1}{8}a[1\bar{1}0]$$

$$p = 0 \qquad q = -\tfrac{1}{8} \qquad q = \tfrac{1}{4} \qquad q = -\tfrac{1}{8}.$$

The foregoing section indicates how field-ion data regarding dissociated dislocations in bcc metals might be interpreted.

The sketches of dislocated surfaces discussed in this section represent only a small proportion of the possible contrast effects which may be expected in field-ion images of bcc metal surfaces, intersected by dissociated dislocations. In each instance the contrast can be predicted in detail, when $p$ and $q$ have been determined, together with the line of intersection of the stacking fault plane and the pole of emergence.

Identification of stacking faults in bcc metals (Bowkett et al., 1969) provides convincing evidence of the utility of the field-ion microscope in dislocation studies of bcc metals and of the validity of the analysis of contrast elaborated here.

Figure 5.24 (cf. fig. 5.20) shows an example of a dislocation which may have dissociated according to the reaction proposed by Crussard (1961) and Cohen et al. (1962). The dislocation emerges in the (121) pole of a tungsten specimen and the line of intersection of the fault plane and (121) is determined directly as [11$\bar{3}$] suggesting a fault on (1$\bar{1}$0), (121) or (211). At an earlier stage in the field evaporation sequence the same dislocation line gave rise to a single spiral in (110). The only $\frac{1}{2}a\langle 111\rangle$ vector which can give a single spiral in (110) and a fault in (121) enclosed by a single spiral is $\frac{1}{2}a[11\bar{1}]$; of the three planes mentioned above a dislocation with this Burgers vector can dissociate only on $(1\bar{1}0)$. The *average* orientation of the dislocation line was determined following the method described in § 1.8.4 as $[14\bar{2}]$. We can

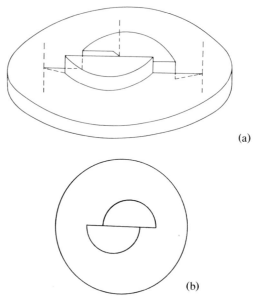

(a)

(b)

Fig. 5.23.   Schematic diagrams of (a) the surface and (b) the contrast expected from one pole of a field-ion tip intersected by a dislocation with $p=0$, dissociated to give two coplanar faults.

therefore conclude that a dislocation has dissociated at *part* of its line in a direction such as $[221]$ as follows:

$$\frac{1}{2}a[11\bar{1}] \rightarrow \frac{1}{8}a[110]_{(1\bar{1}0)} + \frac{1}{4}a[11\bar{2}]_{(1\bar{1}0)} + \frac{1}{8}a[110]_{(1\bar{1}0)}$$
$$p = 1 \qquad q = \frac{3}{8} \qquad\qquad q = \frac{1}{4} \qquad\qquad q = \frac{3}{8}.$$

Since precise $q$ values cannot be deduced from field-ion data, it is sometimes appropriate, e.g. for dislocations in bcc metals (Vitek and Kroupa, 1969) to write dislocation reactions in general form. Then, the Cohen-Crussard reaction for a dislocation emerging in (121) becomes as follows, giving the

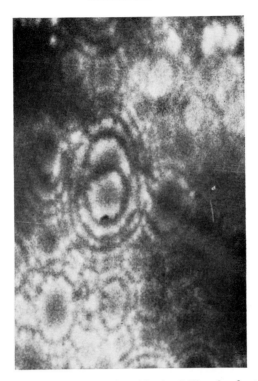

Fig. 5.24. An example of the contrast found in the (121) pole of a tungsten specimen intersected by a dislocation dissociated according to the Cohen-Crussard reaction; cf. fig. 5.20.

quoted $q$ values:

$$\tfrac{1}{2}a\,[11\bar{1}] = \tfrac{1}{2}a\,[m, m, \bar{n}]_{(1\bar{1}0)} + \tfrac{1}{2}a\,[1-2m, 1-2m, \overline{1-2n}]_{(1\bar{1}0)}$$
$$+ \tfrac{1}{2}a\,[m, m, \bar{n}]_{(1\bar{1}0)}$$

$$p = 1 \qquad q = \frac{3m-n}{2} \qquad q = \frac{2-6m+2n}{2} \qquad q = \frac{3m-n}{2}$$

The contrast of fig. 5.24 is also consistent with the following dislocation reaction of Wasilewski (1965) as modified by Foxall et al. (1967).

$$\tfrac{1}{2}a\,[11\bar{1}] \rightarrow \tfrac{1}{6}a\,[11\bar{1}]_{(1\bar{1}0)} + \tfrac{1}{6}a\,[11\bar{1}]_{(1\bar{1}0)} + \tfrac{1}{6}a\,[11\bar{1}]_{(1\bar{1}0)}$$
$$p = 1 \qquad q = \tfrac{1}{3} \qquad q = \tfrac{1}{3} \qquad q = \tfrac{1}{3},$$

where the $p$ and $q$ values again refer to (121). The field-ion microscopic contrast from this dissociation is clearly very similar to that from the Cohen

Crussard reaction since the $q$ values are so close (both reactions are particular cases of the above generalised reaction). A calculation of the difference in expected ledge radii does not permit a distinction of the various possibilities since the $q$ values are all so similar.

Figure 5.25 shows an actual example of a dislocation which has dissociated to give a composite dislocation (Foxall et al., 1967). It will be observed that different planes are defined by the semicircle of bright contrast between the points on the micrograph corresponding to $Y$ and $Z$, fig. 5.21, and the steps in the spiral between those points corresponding to $X$ and $Y$. The ends of the semicircle define a line close to the $[\bar{1}13]$ direction whilst the steps in the anticlockwise stepped spiral define a line close to $[\bar{1}11]$ indicating the fault planes to be $(1\bar{2}1)$, $(2\bar{1}1)$ and $(01\bar{1})$ $(101)$ or $(110)$ respectively. When

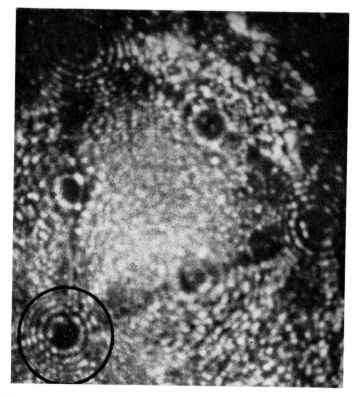

Fig. 5.25. An example of the contrast found in the (110) pole of an iron specimen (imaged with hydrogen at 78 °K) intersected by a composite dislocation; cf. fig. 5.21.
(Courtesy J. Gallot.)

the faults are enclosed by what was a ring of atoms a single spiral is apparent, i.e. $p = 1$ for the parent total dislocation. The possible Burgers vectors for the parent total dislocation are $\frac{1}{2}a[111]$ and $\frac{1}{2}a[11\bar{1}]$.

Self consistent sets of fault planes and Burgers vectors are $\frac{1}{2}a[111]$ dissociated on $(1\bar{2}1)$ and $(01\bar{1})$ or $\frac{1}{2}a[11\bar{1}]$ dissociated on $(2\bar{1}1)$ and $(101)$. The actual reaction could be:

$$\frac{1}{2}a[111] \rightarrow \frac{1}{6}a[111]_{(1\bar{2}1)} + \frac{1}{6}a[111]_{[111]} + \frac{1}{6}a[111]_{(01\bar{1})}$$
$$p = 1 \qquad q = \tfrac{1}{3} \qquad\quad q = \tfrac{1}{3} \qquad\quad q = \tfrac{1}{3}$$

or

$$\frac{1}{2}a[111] \rightarrow \frac{1}{6}a[111]_{(1\bar{2}1)} + \left\{ \begin{array}{c} \frac{1}{6}a[211] \\ \frac{1}{24}a[011] \end{array} \right\} + \frac{1}{8}a[011]_{(01\bar{1})}$$
$$q = \tfrac{1}{3} \qquad\qquad q = \tfrac{13}{24} \qquad\quad q = \tfrac{1}{8}.$$

The fault on $\{112\}$ is between one and two times as wide as that on $\{110\}$.

Theoretical predictions suggest that the dislocation core in bcc metals is extended only one or two Burgers vectors, e.g. Hirsch (1968), Chang (1967). However, field-ion microscopic studies have revealed stacking fault ribbon widths of about 100 Å. This effect may be an artefact of the technique resulting from the shear stresses in a field-ion tip. Alternatively surface relaxations and the large surface to volume ratio, can alter the elastic energy of defects. Also the dilatation of the structure caused by the field stress may alter the stacking fault energy.

## 5.7. Dislocations in hexagonal metals

Field-ion microscopy of the hexagonal metals in general, and the study of dislocations in particular, is complicated by the presence of two non-equivalent sets of atoms, those at 0, 0, 0 and those at $\frac{1}{3}$, $\frac{2}{3}$, $\frac{1}{2}$ in the unit cell. The contribution to the image of the two sets of atoms experimentally is found to depend on the specimen material (Melmed and Klein, 1966, and Melmed, 1966a and 1967) and can be given a physical basis following computer simulations by Perry and Brandon (1969).

Equation (5.1) can be used to predict the multiplicity of dislocation spiral expected in field-ion micrographs of the hexagonal metals providing 3 figure Miller-Bravais indices are used. When $2h + 4k + 3l$ is not a multiple of six, $2h$, $2k$ and $2l$ must be used in eq. (5.1). Although this rule is correct as regards lattice geometry it relies on all atoms contributing to the field-ion image. Brandon and Perry consider this point in more detail. Ranganathan and Melmed (1966) show that in the full four figure Miller-Bravais indices

TABLE 5.4

Equivalence of the Miller and Miller-Bravais indexing systems in calculating $p$ values

| Burgers vector | | Multiplicity of spiral in: | | | |
|---|---|---|---|---|---|
| Miller-Bravais indices 4 figure | 3 figure | $(0001):(001)$ | $(11\bar{2}0):(110)$ | $(10\bar{1}0):(100)$ | $(10\bar{1}2):(102)$ |
| $\frac{1}{3}[11\bar{2}0]$ | [110] | 0 | 2 | 2 | 2 |
| $\frac{1}{3}[2\bar{1}\bar{1}0]$ | [100] | 0 | 1 | 2 | 2 |
| $\frac{1}{3}[\bar{1}2\bar{1}0]$ | [010] | 0 | 1 | 0 | 0 |
| $[0001]$ | [001] | 2 | 0 | 0 | 4 |
| $\frac{1}{3}[11\bar{2}3]$ | [111] | 2 | 2 | 2 | 6 |
| $\frac{1}{3}[2\bar{1}\bar{1}3]$ | [101] | 2 | 1 | 2 | 6 |
| $\frac{1}{3}[\bar{1}2\bar{1}3]$ | [011] | 2 | 1 | 0 | 4 |
| $\frac{1}{3}[\bar{1}\bar{1}23]$ | $[\bar{1}\bar{1}1]$ | 2 | $-2$ | $-2$ | 2 |
| $\frac{1}{3}[\bar{2}113]$ | $[\bar{1}01]$ | 2 | $-1$ | $-2$ | 2 |
| $\frac{1}{3}[1\bar{2}13]$ | $[0\bar{1}1]$ | 2 | $-1$ | 0 | 4 |

Note that whilst there is a very simple relation between three and four figure Miller-Bravais indices for *planes* i.e. if the three figure indices are $(hkl)$ the four figure indices are $(hkil)$ where $i = -(h + k)$, this is *not* the case for the indices of directions.
Partial dislocation reactions which can be found by reference to the review of Partridge (1968) can be analysed as for fcc.

eq. (5.1) becomes

$$hH + kK + iI + lL = p \quad \text{or} \quad \boldsymbol{g} \cdot \boldsymbol{b} \qquad (5.5)$$

for a dislocation with Burgers vector, $\boldsymbol{b}, = [HKIL]$ emerging in the $(hkil)$ plane of unit normal $\boldsymbol{g}$. Figure 5.26 shows a dislocation emerging in $(10\bar{1}0)$ of a ruthenium specimen. Application of eq. (5.1) or (5.5) shows that the Burgers vectors $\frac{1}{3}[2\bar{1}\bar{1}0]$, $\frac{1}{3}[11\bar{2}0]$, $\frac{1}{3}[11\bar{2}3]$, $\frac{1}{3}[2\bar{1}\bar{1}3]$ can generate the observed anti-clockwise two leafed spiral.

The contrast theory developed in this chapter is perfectly general and can be simply extended to any lattice by substitution of the appropriate quantities in eqs. (5.1), (5.2), (5.3) and (5.5). Similar considerations will govern the contrast from any defect describable by a shear, e.g., anti-phase domain boundaries in order-disorder systems. The theory neglects secondary effects due to field evaporation, elastic distortions, and distortion of the ion trajectories by an emergent defect. Field evaporation effects can give a misleading illusion of lattice distortion (Fortes and Ralph, 1969a).

## 5.8. Summary

It is possible to recognise and in some situations characterise perfect dislocations occurring as single dislocation lines or arrays e.g. dipoles.

Fig. 5.26. A dislocation emerging in the (10$\bar{1}$0) pole of a ruthenium specimen. (Courtesy A. J. Melmed.) The (10$\bar{1}$0) pole is in the centre of the picture and the second leaf of the spiral is only partially visible.

Stacking faults can be identified and in particular small scale dislocation dissociations are detectable. The stacking fault ribbon need only be 20 Å wide to exhibit characteristic contrast. Crude estimates of stacking fault energy can be made from field-ion data.

Although artefacts such as Shockley and perfect dislocation loops may be generated by the applied field stress the field-ion microscope can give information, about individual defects, which cannot be obtained by other means. Furthermore it is possible to get information about dislocation behaviour under the influence of complex stresses.

# 6 | INTERFACES

## 6.1. Introduction

It is well known that grain boundaries exert a strong influence on the mechanical properties of materials. In particular grain boundary topography and the presence of a second phase are thought to be major factors. Features at boundaries such as dislocations, protrusions or precipitates may have large effects on properties even when they are irresolvable by conventional transmission electron microscopy. The field-ion microscope has been shown to be able to resolve such details. A contribution can be made to the understanding of such diverse phenomena as creep, intergranular fracture and the influence of grain boundaries on the pattern of radiation damage.

## 6.2. Determination of grain boundary parameters

### 6.2.1. Axis and angle of rotation

The orientation relation between two grains can be described in terms of an axis and angle of rotation which can be estimated as described below. This relative rotation of the two grains across a boundary is of importance in determining the interfacial energy and understanding such effects as grain boundary migration.

The presence of a grain boundary is shown by a breakdown of the usual crystal symmetry; field-ion poles are rotated away from their usual angular positions. A simple graphical method for determining the misorientation employs the assumption that the image is a stereographic projection. This introduces an uncertainty at *least* as big as that normally associated with

142

stereographic manipulations (2°); the method to be described is not as accurate as might be desired but it does provide a rapid way of estimating the grain boundary misorientation. The stereographic constructions used are similar to those employed in the analysis of the orientation of a bicrystal from Laue X-ray data (Goux, 1961 and 1963).

The approximate position of the grain boundary is determined by inspection of the field-ion micrograph. Three pairs of field-ion poles having the same indices with one pole of each kind in grain A and one in grain B are selected, or constructed if necessary. By noting the linear separation on the field-ion micrograph of poles of known angular separation in the *same* grain an average scaling factor can be deduced enabling the plotting of the three pairs of poles on a projection of standard size (preferably 30 cm). Three great circles are constructed, one equally inclined to each of the three pairs of poles. This construction can be done, with the aid of a Wulff net, as follows. Draw

Fig. 6.1.   Grain boundary intersecting a tungsten specimen: 21 °K, helium image gas. Image recorded by fibre-optic technique. Fig. 6.2. is a stereographic projection of this specimen. (Courtesy J. Hren.)

the great circle through $(hkl)_A$ and $(hkl)_B$ and mark the point equally inclined to $(hkl)_A$ and $(hkl)_B$; construct the pole of this great circle. These two points are equally inclined to $(hkl)_A$ and $(hkl)_B$ and permit the construction of the great circle representing the locus of all points equally inclined to $(hkl)_A$ and $(hkl)_B$. The axis of rotation, M, is that direction equally inclined to all pairs of $(hkl)_A$ and $(hkl)_B$ and is defined by the intersection of the three great circles of equal inclination. The relative rotation $\alpha$ of grains A and B can be measured directly on the stereogram by standard means (i.e. the angle between the tangents at M to the great circles through M and $(hkl)_A$ and M and $(hkl)_B$) or more accurately by determining the angle between the poles of these two great circles. The above method applied to the grain boundary in fig. 6.1 showed that the crystals were related by a rotation of approximately $31°$ about an axis just off the [110] of either grain.

### 6.2.2. GRAIN BOUNDARY PLANE OR PLANES

Determination of a grain boundary plane necessitates the use of several

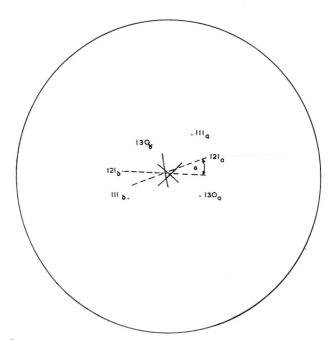

Fig. 6.2. Stereogram used to analyse the boundary shown in fig. 6.1. Only segments of the great circles of equal inclination are marked. The dashed lines are tangents, at the axis of rotation, to the great circles between $121_a$ and the axis and $121_b$ and the axis.

approximations which combine to make the procedure rather inaccurate. The idealised problem which was considered in § 1.8.5 of the analysis of the intersection of a plane with a spherical surface projected stereographically is comparatively simple; with appropriate scaling, the trace can be transferred to a standard stereogram. In general the trace will be a small circle and its pole may be constructed on the stereogram. In practice this procedure may be far from simple because the tip is not always a spherical cap, the projection is not stereographic and grain boundaries are not always planar over the tip diameter. Nevertheless the above method can be applied to small lengths of grain boundary trace in regions of constant local radius. Another method is to calculate the inclination of the boundary to the specimen axis by noting the movement of the trace of the grain boundary during a known amount of field evaporation. This can be done using a formula developed in § 1.8.4. for estimating the inclination of a line defect to the specimen axis. Morgan and Ralph (1968) have elaborated this approach to take account of different end-forms. When the grain boundary trace is obviously segmented an analysis can be performed on each small region of different orientation.

### 6.3. Low angle grain boundaries

The dislocation theory of the structure of low angle boundaries first described by Read and Shockley (1950) has been well substantiated by optical, X-ray and electron microscopical techniques. However some doubt remains as to the magnitude of misorientation at which the theory breaks down. There is great interest in the possibility that dislocations which may be present in high angle grain boundaries could account for grain boundary migration and sliding by their movement. It is the purpose of the following sections to illustrate some extensions of the contrast theory of the field-ion microscope and to indicate how field-ion microscopic study of grain boundaries may be expected to yield new information about their structure.

6.3.1. LOW ANGLE GRAIN BOUNDARIES: CONTRAST THEORY

It will be recalled that the contrast from a single perfect dislocation emerging in the pole with normal $n$ is $p$ interleaved spirals where $p$ is the resolved component of $b$ in $n$, measured in units of the appropriate interplanar spacing (§ 5.2.1). When more than one dislocation is enclosed by an imaged chain of atoms the spiral configuration depends on the *sum* of the Burgers vectors enclosed and spirals of unexpected multiplicity can be produced.

This result follows from the property that a ring of atoms on the specimen surface constitutes a Burgers circuit.

Possible dislocation descriptions of a low angle grain boundary can be deduced from Frank's formula (Read, 1953; Amelinckx and Dekeyser, 1959):

$$S = (u \times V)\theta$$

where $S$ is the closure failure of a Burgers circuit, $u$ is the axis of misorientation, $V$ is a vector in the plane of the boundary and $\theta$ is the relative rotation of the adjacent grains. The contrast of the field-ion image can then be predicted simply, but shows a number of features which are worthy of mention. As an example consider a pure twist boundary on (111) in an fcc metal. Use of Frank's formula indicates that the boundary may be made up of an hexagonal grid of screw dislocations having Burgers vectors $\frac{1}{2}a[\bar{1}10]$, $\frac{1}{2}a[01\bar{1}]$ and $\frac{1}{2}a[10\bar{1}]$. Supposing the boundary happens to intersect the field-ion tip surface in the [111] zone the multiplicity of spiral expected is as shown in table 6.1.

TABLE 6.1

Multiplicity of spiral expected from the dislocations* in a twist boundary on (111), intersecting a tip in some poles of the [111] zone

| $(hkl)$ <br> $b$ | $2\bar{2}0$ | $0\bar{2}2$ | $\bar{2}\bar{2}4$ | $\bar{2}02$ | $\bar{2}20$ |
|---|---|---|---|---|---|
| $\frac{1}{2}a[10\bar{1}]$ | 1 | $-1$ | $-3$ | $-2$ | $-1$ |
| $\frac{1}{2}a[01\bar{1}]$ | $-1$ | $-2$ | $-3$ | $-1$ | 1 |
| $\frac{1}{2}a[\bar{1}10]$ | $-2$ | $-1$ | 0 | 1 | 2 |

* For the purposes of the analysis of field-ion micrographs, Burgers vectors are not defined with positive line sense looking out from the node, but with a common line sense; this means that the sum of the Burgers vectors will not be zero.

The spacing of parallel dislocations is about $b/\theta$ for each set i.e. about 40 Å for a 3° boundary. This result implies that for most tip radii more than one dislocation would emerge in the central ring. In planes such as (0$\bar{2}$2) where $p$ is negative for all dislocations, high order multiple spirals may be produced; conversely in planes such as ($\bar{2}$20) where $p$ can take either sign, a contrast effect which could be mistaken for a dislocation loop can result. A further "artefact" effect is that owing to the nature of the image projection equally spaced dislocations would not immediately appear to be equally spaced in a micrograph. Figure 6.3 shows the striking contrast we might expect to see in the ($\bar{2}$20) pole where the $p$ values for the dislocations have opposite signs and different magnitudes.

Ryan and Suiter (1966) show an example of a multiple spiral configuration in the (110) pole of a tungsten field-ion specimen intersected by a twist boundary; see fig. 6.4. The triple spiral effect could be ascribed to three $\frac{1}{2}a$[111] dislocations having $p = 1$ rather than a single dislocation of Burgers vector $\frac{3}{2}a$[111] etc. Some other reported observations of low angle boundaries suggest that preferential field evaporation of strained atoms gives a major contribution to the image contrast. This is interesting since preferential field evaporation does not appear to be a noticeable effect around single dislocations.

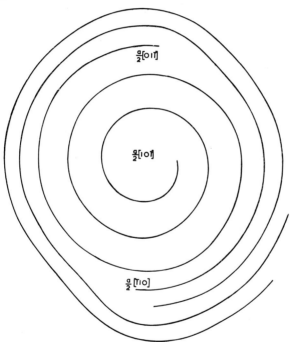

Fig. 6.3.    Diagram illustrating the type of multiple spiral effect produced by a sub-boundary emerging in a field-ion tip.

### 6.3.2. EXAMPLES OF LOW ANGLE GRAIN BOUNDARIES

Figure 6.5 shows an example of a low angle grain boundary in an iridium field-ion specimen. The boundary plane trace remained approximately as in fig. 6.5 during field evaporation, from which it can be concluded that the indices of the boundary plane are close to (1$\bar{1}$0). The rotation of the two grains

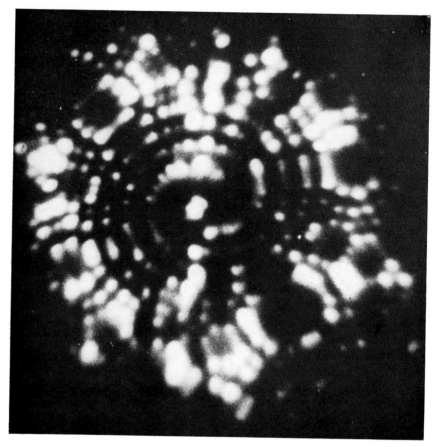

Fig. 6.4.   Example of multiple spirals produced in the (110) pole of a tungsten specimen
containing a low angle boundary (78 °K helium image gas). (Courtesy J. Suiter.)

can be regarded as $10°$ about $[001]$ and about $3°$ about $[1\bar{1}0]$, (the misorientation is really a single rotation about an axis between $[001]$ and $[1\bar{1}0]$,
but it is more convenient for the present purpose to regard it as the sum
effect of two rotations). Frank's formula (6.1) shows that the dislocations
generating the former rotation give $p=0$ and hence are not responsible for
the spiral contrast in (001). However the dislocations generating the rotation
about $[1\bar{1}0]$ have $\boldsymbol{b}=\frac{1}{2}a[01\bar{1}]$, $\frac{1}{2}a[011]$ or $\frac{1}{2}a[\bar{1}\bar{1}0]$ the first two of which
have $p\neq0$. The dislocation spacing is predicted to be about 55 Å and indeed
five spirals can be seen to begin between (113) and $(\bar{1}\bar{1}3)$ which are separated

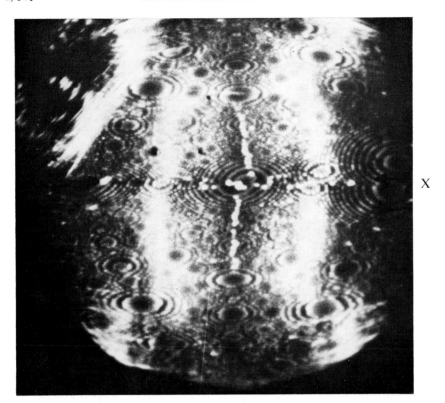

X

Fig. 6.5.   Low angle boundary in an iridium specimen (78 °K, helium image gas).
(Courtesy M. A. Fortes.)

by about 250 Å in the micrograph of fig. 6.5. Spiral contrast can also be seen in the adjacent {111} planes near the point X. Figure 6.6 shows an example of a sub-boundary in tungsten. The axis of rotation $u$ is near $[1\bar{1}0]$ and the boundary plane was shown to be close to $(1\bar{1}0)$ in the same way as for the previous grain boundary. The boundary can be described by an array of dislocations of Burgers vectors $a[100]$ and $a[0\bar{1}0]$. These two vectors give $p = \pm 1$ in (110) respectively. An example of dislocations giving spirals of opposite sense (implying opposite $p$ values) is visible in the central region of (110). Many other spirals can be seen in (110).

Both the boundaries shown in figs. 6.5 and 6.6 are predominantly "twist". However, careful inspection of the manner in which the planes meet, reveals that the boundaries are in fact of mixed character (§ 6.4.1).

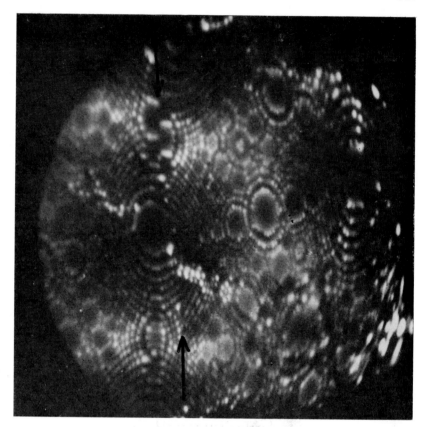

Fig. 6.6.   Low angle boundary in a tungsten specimen (78 °K, helium image gas).

## 6.4. High angle grain boundaries

### 6.4.1. STRUCTURE

Understanding of the structure and misorientation of high angle grain boundaries has been clarified considerably by field-ion microscopic observations. The Mott (1948) "island" model and the coincidence site lattice theory, with additions, have been demonstrated to be in agreement (within the accuracy of the technique) with observations of the grain boundary structure in recrystallised metals (Brandon et al., 1964; Ranganathan, 1965; Brandon, 1966e). It is difficult to deduce the atom arrangement at a grain boundary; rather the intuitive correlation between misfit and high energy leading to preferential field evaporation, has sometimes been invoked to indicate bad fit. Where the fit is good little preferential field evaporation is expected on this model and conversely where fit is bad a dark band analogous to an etch pit will be found adjacent to and at the boundary. Other effects may play a part. The regions of good fit, within experimental uncertainty, correspond to densely packed planes in the coincidence site lattice. The direct resolution of the atom arrangement at a grain boundary requires low specimen temperatures. Field-ion evidence suggests that grain boundaries are only two atoms or so wide but it is difficult to state the location of atoms in a field-ion tip sufficiently precisely. By analogy with the conventional model for the width of a dislocation we shall define the "width" of a grain boundary as the distance between points where atoms are displaced from their equilibrium position more than one quarter of the shortest vector of the perfect lattice i.e. it is necessary to measure displacements of only 0.5 Å or so.

Ryan and Suiter (1964) have obtained micrographs (see fig. 6.7) of grain boundary protrusions such as those that have been postulated in mechanisms of the initiation of creep failure.

Although there have been a number of papers reporting field-ion observations of grain boundary orientation and topology few attempts have been made to understand the contrast effects resulting from the way in which planes of atoms meet across a grain boundary. The imaged rings

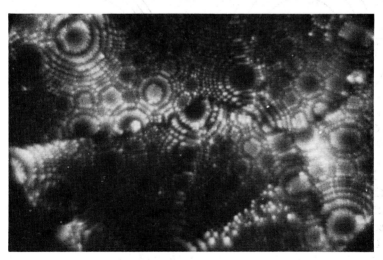

Fig. 6.7.   Grain boundary protrusions in a tungsten specimen (78 °K, helium image gas).
(Courtesy J. Suiter.)

of atoms adjacent to a high angle grain boundary adopt configurations governed by the local tip shape, the misorientation and also the stage that field evaporation has reached. The net result of these various factors is difficult to quantify in a general form but it can be shown how spiral structures having a misleading resemblance to dislocations and stacking faults can be produced.

Consider a grain boundary in a field-ion tip of a cubic metal (although the ideas have general validity); if the direction $[hkl]_A$ is parallel to $[hkl]_B$ and the planes $(hkl)_A$ and $(hkl)_B$ are at the same level, then the image is perfect in the $(hkl)_A/(hkl)_B$ region except for displacements which may be irresolvable at the grain boundary. However, if $(hkl)_A$ and $(hkl)_B$ are at different levels (i.e. there is a mismatch along $[hkl]_{A \text{ and } B} \neq nd_{hkl}$ where $n$ is an integer) then the contrast observed will resemble that from a stacking fault; distinction is possible by noting the misorientation of planes other than $(hkl)$. More generally, $[hkl]_A$ will not be parallel to $[hkl]_B$ then the $(hkl)$ ledges will not match perfectly whether or not $(hkl)_A$ and $(hkl)_B$ are at the same level. An enormous variety of structures are possible and two are illustrated in fig. 6.8. Similar effects can arise when the Miller indices $(hkl)_A$ and $(h'k'l')_B$ of the two adjacent planes are not the same.

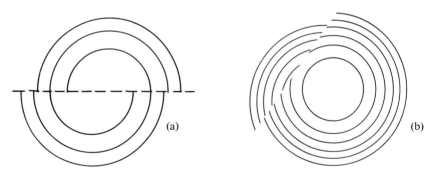

Fig. 6.8.    Diagrams of structures expected in the field-ion image because of a "twist" type rotation of one grain with respect to another. (Courtesy M. A. Fortes.)

Planes of field-ion poles which are neither perpendicular nor parallel to the axis of rotation give ring patterns as illustrated in fig. 6.9. This effect follows from the field evaporated endform's dual boundary conditions: (i) smoothly curved evelope, (ii) low index regions develop as well defined facets. It should be emphasised that the matching of the adjacent lattices can be good (except at dislocation spirals) despite the absence of regular rings. Grain boundaries must be considered individually, and we have only outlined

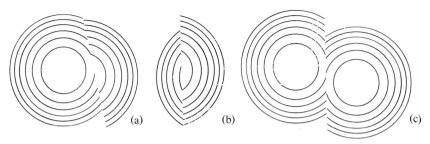

Fig. 6.9.   Diagrams of structures expected for a general rotation (see text).

some of the more obvious aspects of the problem. Fortes and Smith (1970) have considered other examples.

### 6.4.2. GRAIN BOUNDARY SEGREGATION

It is known that very small amounts of solute can drastically reduce the ductility of metals, possibly by forming a monatomic grain boundary layer or very small grain boundary precipitates. The field-ion microscope is apparently an ideal tool for studying grain boundary segregation. There are however two important difficulties in using the field-ion microscope:
  (i) it is necessary to be able to differentiate between matrix and solute,
 (ii) it may not be easy to study equilibrium effects since the grain size after an anneal may be too large for a useful probability to remain that a grain boundary will be observed in a tip.

A solution to the first problem may be to use the 'atom probe' to establish the imaging behaviour of solute unambiguously in conjunction with the theories of alloy field evaporation considered in ch. 9. The second problem may be approached in a number of ways e.g. electron microscopic selection of tips before field-ion imaging or by recrystallising heavily cold worked material under conditions where the solute simultaneously diffuses into the matrix.

Despite the difficulties Fortes and Ralph (1967) have shown that it is feasible to study grain boundary segregation for a system giving unmistakeable contrast effects e.g. oxygen in iridium (see ch. 9) and to plot segregation profiles.

## 6.5.  Twins

Twins can readily be recognised in the field-ion image since they appear as planar interfaces across which there is a displacement or appropriate

Fig. 6.10.    Twin lamella in an iridium specimen (21 °K, helium image gas).
(Courtesy E. W. Müller.)

lattice rotation. Twin lamellae have been observed frequently in "bulk" material and electrodeposited layers (fig. 6.10). The curvature of the boundary in the image is what one expects from a planar intersection with an approximately hemispherical field-ion tip. Note that the *curved* intersection through the *centre* of the image immediately shows that the defect trace is not a great circle. That the lamella is in a twin orientation can be deduced from the standard grain boundary analysis described in § 6.2.1, or simply by noting the indices of the coincident poles in the twin and matrix. Such twins, which are formed *in situ* are most common in specimens with a ⟨111⟩ axis (Rendulic and Müller, 1966; Rendulic, 1967; Fortes and Ralph, 1969b).

Bicrystals in twin orientation have been seen in field-ion images of tungsten (Hren, 1965). Very narrow microtwins have been observed in W-5% Re deformed in tension at 78 °K. Such defects give streak contrast at the interface owing to the displacement $\frac{1}{6}a\langle 111 \rangle$ which in general yields a step of height a non-integral number of planar spacings.

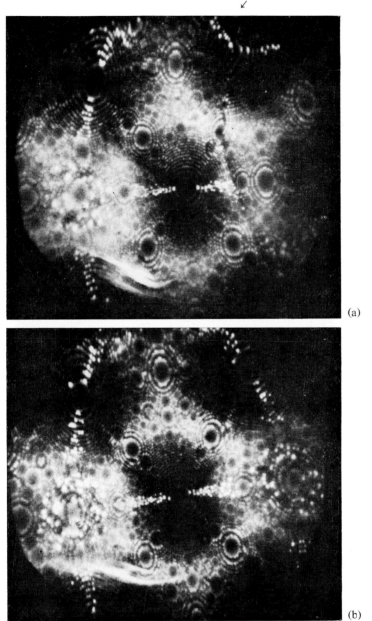

(a)

(b)

Fig. 6.11.   Micrographs of a tungsten specimen showing an untwinning event caused by
the field stress (a) twinned (b) two planes evaporated: no twin visible
(78°K, helium image gas).

It is interesting to note that a coherent twin boundary will not generate any spiral configurations. In an incoherent twin boundary dislocations are incorporated and their presence generates spiral contrast which should be one of the easiest cases for recognising dislocations in a high angle boundary.

Just as deformation twins are formed *in situ* by the operation of the field stress, it has been observed that untwinning can also occur. Figure 6.11 shows the image of a tungsten specimen which initially was cut by a twin boundary which was eliminated when the field (and hence stress) was raised for field evaporation.

### 6.6. Domain boundaries

There are two broad classes of domain boundaries in ordered alloys: rotational and translational. A rotational domain boundary can be detected as a mismatch and rotation and analysed in a way similar to that described for high angle grain boundaries.

A translational domain boundary (see fig. 9.1) marks the slip plane of a partial dislocation of the superlattice. The contrast expected is precisely analogous to that from a stacking fault and the visibility criteria are the same. Contrast is seen because only one species contributes to the image (Southworth and Ralph, 1966; Tsong and Müller, 1966a). The field-ion microscope has great potential in the study of dislocation configurations in ordered alloys and may indicate the physical reality of some of the dislocations postulated by Marcinkowski (1964) and Reudl et al. (1968). The contrast from ordered alloys is considered further in ch. 9.

### 6.7. Coherent and partially coherent precipitate-matrix interfaces

It is possible to define a Burgers vector for a precipitate by considering a Burgers circuit similar to that used for defining the Burgers vector of a stacking fault (Kelly and Nicholson, 1963). A coherent precipitate has a Burgers vector less than the shortest lattice vector, in contrast to a partially coherent precipitate for which the Burgers vector is greater than the shortest lattice vector. As a consequence chains of matrix atoms which would cross a precipitate show a closure failure in a similar fashion to chains of atoms which enclose a dislocation line. An interesting example of a misfit dislocation in a partially coherent interface is shown in fig. 6.12. A layer of molybdenum ($a = 3.14$ Å) has been evaporated onto a field evaporated tungsten substrate ($a = 3.16$ Å); field

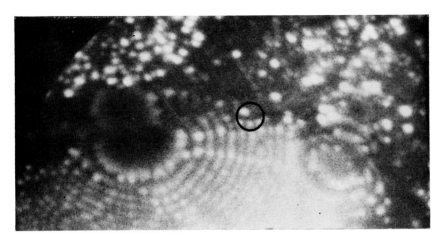

Fig. 6.12.   Dislocation in a partially coherent tungsten-molybdenum interface (78 °K, helium image gas). (Courtesy E. Boyes, The General Electric Company Ltd.)

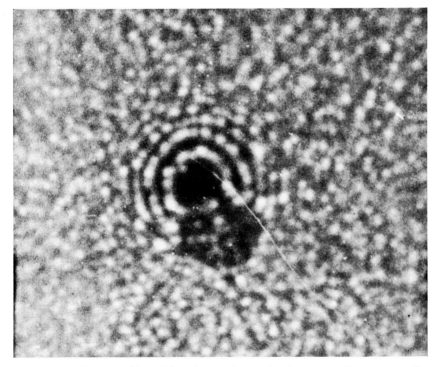

Fig. 6.13.   Coherent Fe₃Si precipitate in a steel; note the rings are continuous across the interface (78 °K hydrogen gas). (Courtesy D. Schwartz.)

evaporation has been used to expose the tungsten-molybdenum interface. As would be expected from their common bcc structure and the small misfit $\delta [= 2(a_1 - a_2)/(a_1 + a_2)]$ the matching of atomic rings across the interface is very good. Since however $a_{Mo} \neq a_W$ an interface dislocation is required at intervals of $a\sqrt{3}/2\delta$ to relieve the accumulated strain due to the misfit. One such dislocation, generating a spiral continuous in the tungsten and molybdenum is marked in fig. 6.12. Clearly the field-ion microscope has great potential in the direct study of the breakdown of coherency.

Figure 6.13 is an image of a coherent $Fe_3Si$ precipitate in a steel: note the continuous rings in precipitate and matrix.

The plane of a platelike precipitate can be deduced from its line of intersection, in general a small circle, and its pole. Small precipitates may be crystallographically characterised by direct comparison of the line of intersection with known zone lines; this method is particularly convenient

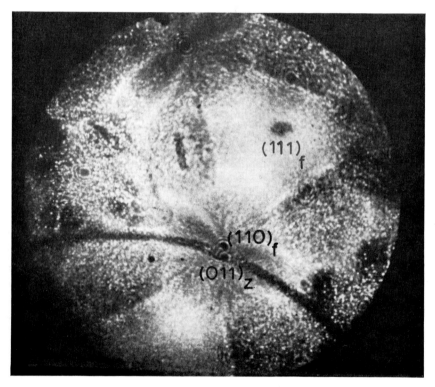

Fig. 6.14. Pearlite showing coincidence of poles in matrix and precipitate (78 °K, hydrogen image gas). (Courtesy R. Morgan.)

if it is simply desired to distinguish a few specific possibilities e.g. a $\{111\}$ from a $\{100\}$ habit.

Morgan (1967) has successfully determined a relationship between directions in precipitate and matrix by noting that if two sets of part rings have the same pole then the normal to $(hkl)$ matrix is parallel to the normal to $(pqr)$ precipitate (fig. 6.14). Another method for determining the plane, applicable to platelike precipitates is to note the local normal at the point where a precipitate disappears during field evaporation. This direction is the normal to the matrix plane which is parallel to the plane of the precipitate.

# 7 | 3-DIMENSIONAL DEFECTS

## 7.1. Introduction

The field evaporation process makes the field-ion microscope particularly suitable for studying 3-dimensional defects. Making use of carefully controlled field evaporation it is possible to determine accurately the size and shape of defects. Furthermore the problems encountered with thin-foil electron microscopy, when defect images may overlap, do not arise. The field-ion microscope has been used to study the debris left behind after irradiation by particles with high momentum, and also voids and fine scale precipitation in a variety of materials.

It is necessary to take many micrographs in order to reconstruct a 3-dimensional defect and it is especially helpful for this type of study to have some form of image intensification to reduce the amount of time spent on photography. In addition a pulse generator can make control of the field evaporation process easier, and also more uniform, than is possible with manual control.

## 7.2. Fission fragment damage

An experimental arrangement for the fission fragment irradiation of specimen tips is shown in fig. 7.1. The size of such a capsule is restricted by the irradiation facilities available in a reactor and the geometry must be chosen to give the best possible collimation of the fission fragments. In fact the angular spread of particles for any one specimen in the assembly illustrated is less than 7°. The field-ion specimens are mounted in this assembly in the same way that they are mounted in the microscope.

A small piece of enriched uranium foil is positioned several millimetres from the specimen tips and when the capsule is in the reactor neutron bombardment induces fission and some of the fission fragments produced are emitted from the surface of the foil. Enriched uranium foil (90% $^{235}$U) should be used in order to minimize neutron damage in the specimens; the capsule needs to be sealed under vacuum because the range of fission fragments in air is comparatively small. The relative intensities of the spectrum of fission fragment energies may be controlled by varying the thickness of the uranium foil (Redmond et al., 1962).

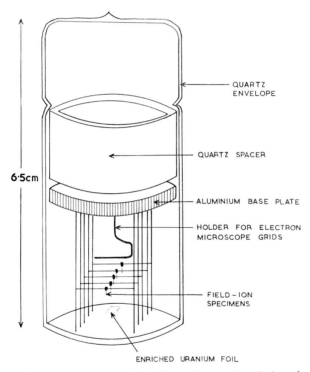

QUARTZ
ENVELOPE

QUARTZ SPACER

6·5cm

ALUMINIUM BASE PLATE

HOLDER FOR ELECTRON
MICROSCOPE GRIDS

FIELD – ION
SPECIMENS

ENRICHED URANIUM FOIL

Fig. 7.1.   Experimental arrangement for the fission fragment irradiation of specimen tips.

In experimental results reported by Bowkett et al. (1965 and 1967) the estimated total neutron dose of the specimens was $5 \times 10^{14}$ nvt. The fission fragment dose was monitored by means of molybdenum trioxide crystals sublimed onto carbon coated microscope grids which were mounted alongside the field-ion specimens as shown in fig. 7.1. A fission fragment passing through a molybdenum trioxide crystal registers as a hole and each individual hole represents the passage of a single particle (Bowden and Chadderton, 1962). Figure 7.2 is an electron micrograph of such a crystal. Ideally one and only one fission fragment should strike each specimen, and this condition was achieved as closely as is possible within the limits of a statistical distribution. In the electron micrograph there is on average one hole in every $3 \times 10^{-11}$ cm$^2$ which is comparable with the area of cross section of a typical field-ion tip.

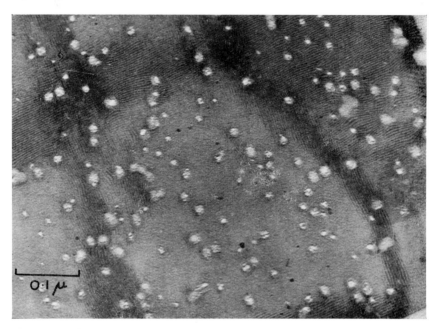

Fig. 7.2.   Molybdenum trioxide crystal used to monitor dose. Each spot marks the point where a fission fragment has hit the crystal (electron micrograph).

Where a fission fragment strikes the surface of a tungsten specimen, the damage observed takes the form of a crater which appears very dramatically in the field-ion image as shown in fig. 7.3. By field evaporation atomic layers may be stripped from the surface and the depth of these cavities can be found directly. The pattern of damage within the specimen may also be investigated. Figure 7.3 is typical of how a specimen first appears once the four to five layers of oxide and adsorbed impurity on the surface have been field evaporated. The very large central crater was about 250 Å radius, whilst the specimen surface itself was about 500 Å radius. The shape of the cavity was found to be approximately hemispherical by field evaporating

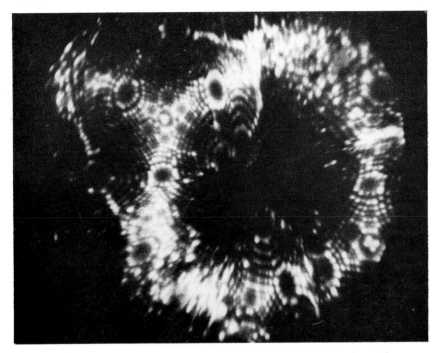

Fig. 7.3.   Field-ion micrograph of a tungsten specimen bombarded by fission fragments showing large surface crater.

the specimen layer by layer and comparing micrographs after evaporation of each layer.

Fission fragments producing cascade events some distance below the surface might be expected to produce vacancy clusters having a size considerably smaller than the cavities at the surface. Such clusters are in fact uncovered by field evaporation. Figure 7.4 shows a section through a cluster of this type which had a diameter of 35 Å. Isolated single vacancies and a few small clusters which were also detected could have been generated by fission fragments or by neutrons, since the specimens also received a neutron dose of $5 \times 10^{14}$ nvt. Separate investigations (e.g. Bowkett et al. 1964) on neutron-irradiated tungsten wire which had received a comparable dose have shown that the clusters due to neutron damage will be nothing like as large as the one shown in fig. 7.4, which can thus be attributed unambiguously to fission fragment damage. Craters of these dimensions were never observed in the tungsten used for these experiments when it was examined either unirradiated or neutron-irradiated; however, very occasionally, poorly

Fig. 7.4.   35 Å diameter sub-surface cluster revealed by field evaporation.


prepared tungsten specimens can give rise to surface holes which bear a superficial resemblance to fission-fragment-induced craters.

Figures 7.5 to 7.8 show a field evaporation sequence through a specimen which had been irradiated in the manner described above. Initially (fig. 7.5) the specimen surface was somewhat damaged round the edges and there was a large region of dim contrast close to the centre of the specimen, indicating the presence of a surface depression. Evaporation revealed the cause as a cavity below the surface and about 50 Å in diameter (fig. 7.6); the presence of a void or large vacancy cluster leads to a weakening effect which may be estimated using the Griffith crack formula (e.g. Cottrell, 1965). For a void of radius $r$ in a material of surface energy $\gamma$ and Young's modulus $E$ the specimen can fracture if:

$$\sigma \geqslant (E\gamma/r)^{\frac{1}{2}}.$$

The field stress for imaging is $\approx 10^{11}$ dyn·cm$^{-2}$. Young's modulus is about $2 \times 10^{12}$ dyn·cm$^{-2}$ and the surface energy is about $5 \times 10^3$ erg·cm$^{-2}$. Fracture is expected for $r \leqslant 10^{-6}$ cm. Such an effect would of course be

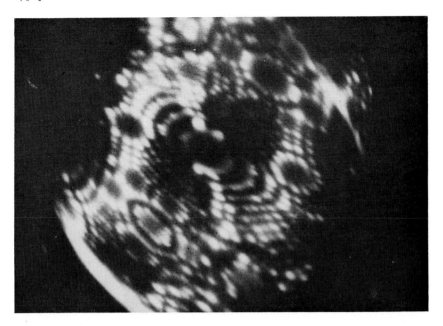

Fig. 7.5. Fission fragment irradiated tungsten specimen. Many layers have been field-evaporated from the surface which still has a somewhat disturbed structure (note especially the dark region on the central (110) plane indicating the presence of a sub-surface vacancy cluster).

magnified in the presence of an excess pressure of gas in the void, and fracture expected at a lower radius.

It is sometimes possible to distinguish in the field-ion microscope between a dispersed region having a high vacancy concentration and a true void (see ch. 4). In this case the cavity was a true void. Examination of field-ion micrographs such as fig. 7.7 showed that the cavity was pear-shaped with a long tail which finally emerged at the side of the specimen (fig. 7.8). This event may be similar to that observed by Bowden and Chadderton (1962) in thin crystals where the track of an individual fission fragment (as distinct from cascade events) showed up in the electron microscope as a light area on a dark background interpreted as representing removal of some of the crystal by the particle.

With non-axial bombardment the fragment track emerges from the side of the specimen and material can be ejected. Naturally this would not be the case in bulk material where atoms knocked off lattice sites must

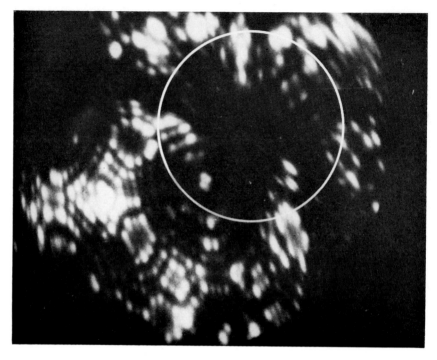

Fig. 7.6.   After the evaporation of two atomic planes from the surface depicted in fig. 7.5, the vacancy cluster is revealed. The position is marked by the white circle, which has a diameter about twice that of the damage region.

occupy interstitial sites, agglomerate or effectively be annihilated on other lattice defects.

In the sequence (figs. 7.5 to 7.8) it can be demonstrated that the fission fragment particle entered the specimen at an angle to the specimen axis from analysis of the translation of the damage region normal to the axis, during field evaporation.

The angle of incidence, $\alpha$, can be determined by plotting out the observed damage in the form of a 3-dimensional model*, as shown schematically in fig. 7.9. The angle $\alpha$ here is found to be $\approx 23°$ which agrees well with the experimental arrangement, where the fission fragments strike the specimen at $20° \pm 7°$ to the axis.

A grain boundary can be seen on fig. 7.8 and it is possible that the

* Because there is an appreciable radius change this method is more appropriate than applying the relation derived in § 1.8.4. to find the angle $\alpha$.

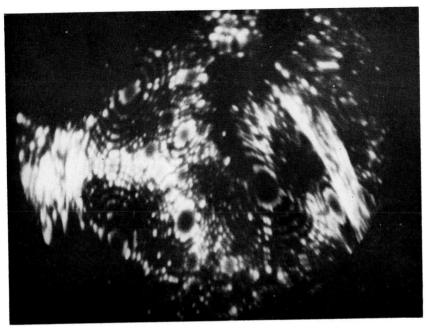

Fig. 7.7.   The same tungsten specimen as fig. 7.6 after evaporating a further 100 Å, the damage region is healing and moving across the specimen surface.

observed damage pattern has been modified by the presence of the boundary.

In view of the evidence for large surface craters associated with the incidence of fission fragments it is worth while considering why such a crater is not seen here. In the work described earlier axial fission fragments were used and under these conditions it is to be expected that much of the particle momentum would be transferred to the specimen to cause removal of surface atoms. In the latter set of experiments the fission fragments will transfer their energy in such a way as to remove atoms from the specimen where the fragment emerges.

This result lends strong support to the concept that in small specimens a column of material is 'pushed out' of the solid along the track. The earlier results are seen as indicating the nature of the 'cascade' regions of fission fragment tracks and are thus more useful in elucidating mechanisms of fission fragment damage in bulk crystals.

The field evaporation process is likely to limit observation of some details of fission fragment damage since any small asperities could be removed at a low voltage when the image would be difficult to observe. Such a problem

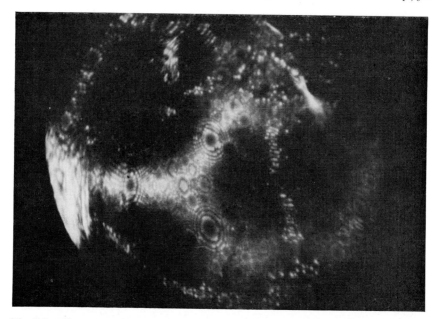

Fig. 7.8. The same as fig. 7.7. after evaporating a further 290 Å. The damage has nearly healed, but there is a grain boundary running across the damaged region (110), with a misorientation of about 12°.

could be alleviated by the use of an intensifying device. A more important difficulty is that a void may be revealed by a fracture process rather than controlled field evaporation and artefacts may be introduced both mechanically and by a modification of the imaging process, e.g. the projection could change for specimens not having a smooth endform. In the case of multiple tips image distortions such as streaks can be expected as a consequence of interaction between ion beams from different regions of the tip. Interpretation of crater shapes should take account of the fact that an image is formed only of those edges of the crater which have a sufficiently small radius of curvature and less drastic changes of topology can only be inferred from field evaporation sequences.

## 7.3. Precipitation

The nature of the contrast to be expected at the precipitate matrix interface has been discussed in § 6.6. In this section we consider what information may be obtained by field-ion microscopic study of precipitation. Although the atomic resolution of the field-ion microscope is well known there are a number of areas where it has found useful application employing a resolution of the order of tens of Ångstrom units. These include situations where the high density and small size of defects, e.g. small precipitates in the early stages of ageing, render the interpretation of electron micrographs difficult, tedious or unreliable.

The detailed treatment of the metallography of pearlite by Morgan and Ralph (1968) showed that orientation relationships for a single precipitate with the matrix could be determined (see § 6.7). The size and distribution of cementite lamellae were deduced from field-ion and electron microscopic

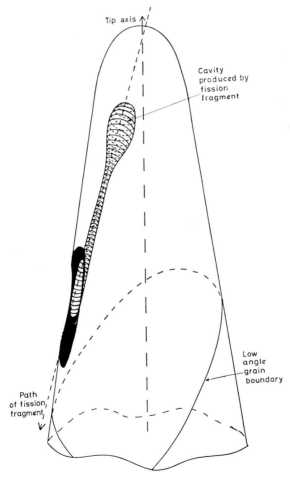

Fig. 7.9.   A schematic representation of the damage induced by the fission fragment as it passed through the crystal.

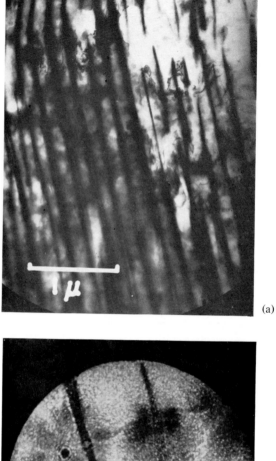

(a)

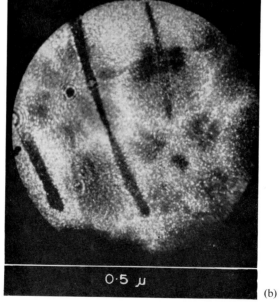

(b)

Fig. 7.10.   Pearlite: (a) electron micrograph and (b) field-ion micrograph, prepared from
the same material. (Courtesy R. Morgan.)

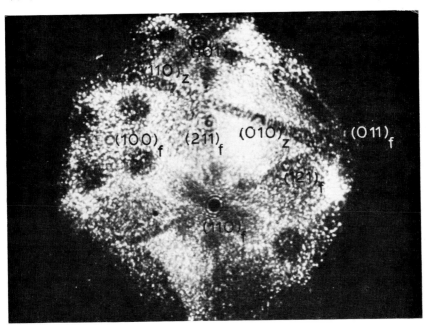

Fig. 7.11.    Indexed field-ion micrograph of pearlite. (Courtesy R. Morgan.)

observations of specimens prepared from the same starting material. Figures 7.10a and b show electron and field-ion micrographs of specimens prepared from similar starting material. The spacing of the cementite lamellae can be seen to be comparable in the two images. Figure 7.11 shows an indexed pearlite micrograph and from many such images the orientation relationships shown in table 7.1 were deduced. These are in good agreement with the average values obtained by other means. It is worth repeating that field-ion microscopy is a convenient way of determining the individual orientation relationship for *each* lamellae and the matrix.

Similarly, Gallot, Honeycombe and Ralph (1967, private communication) and Schwartz et al. (1968) have studied precipitation in secondary hardening steels based on Fe-Mo-C and Fe-V-C respectively by field-ion

TABLE 7.1
Orientation relationship between cementite and ferrite in pearlite (Morgan, 1967)

| $(100)_z$ | $\parallel$ | $(\bar{1}10)_f$ | $z$ = cementite |
| $(010)_z$ | $\parallel$ | $(111)_f$ | $f$ = ferrite |
| $(001)_z$ | $\parallel$ | $(11\bar{2})_f$ | |

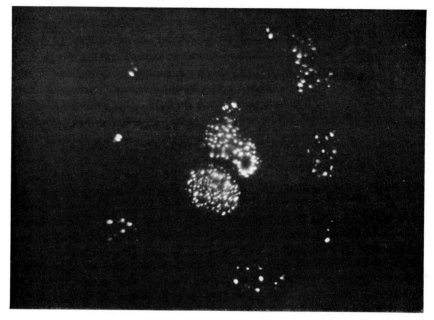

Fig. 7.12a

and electron microscopy. Field-ion microscopy extends the range of metallographic phenomena which may be studied, whilst the integrity of the field-ion observations is assured by comparison of coarser scale details with electron micrographs.

A major difficulty in the studies mentioned above has been that stable atomically resolved images of precipitate and matrix, have not yet been obtained. The problem in these systems is that the matrix and precipitate do not have the same $F_E$ but are at the same potential.

Once the field evaporated endform has been established the boundary conditions are that the tip is an equipotential, and that the matrix and the precipate field evaporate simultaneously. At the evaporation voltage, $V_E$, both the matrix and precipitate are at $F_E$ but in general $F_E$ for the matrix $(F_E^m)$ does not equal $F_E$ for the precipitate $(F_E^p)$. We assume that the precipitate and matrix are such as to obey the same law between field and applied voltage, then:

$$F_E^p = \frac{V_E}{kR_p} \qquad (7.1)$$

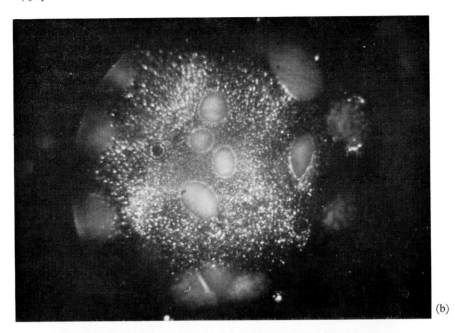

(b)

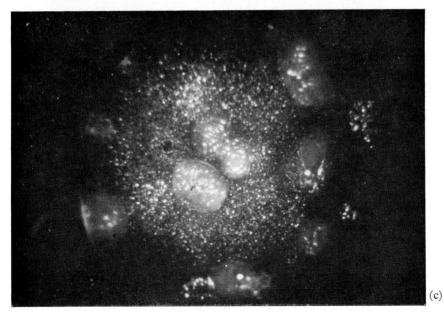

(c)

Fig. 7.12.   Field-ion microscopy of an Fe-Mo-C alloy; (a) shows a hydrogen ion image
of Mo$_2$C, (b) shows the matrix with the precipitate above best image field and blurred,
(c) the result obtained using a hydrogen/neon gas mixture. (All at 78°K.) (Courtesy
J. Gallot and P. J. Turner.)

and

$$F_E^m = \frac{V_E}{kR_m}$$     (7.2)

where $R_p$ and $R_m$ are the local tip radii for precipitate and matrix. Hence:

$$\frac{F_p}{F_m} = \frac{R_m}{R_p}$$     (7.3)

e.g. for Fe-Mo$_2$C where $F_E^{Fe} \simeq 3v/\text{Å}$ and $F_E^{Mo_2C} \simeq 4$ V/Å (say), then:

$$\frac{R_{Fe}}{R_{Mo_2C}} = \frac{4}{3}.$$

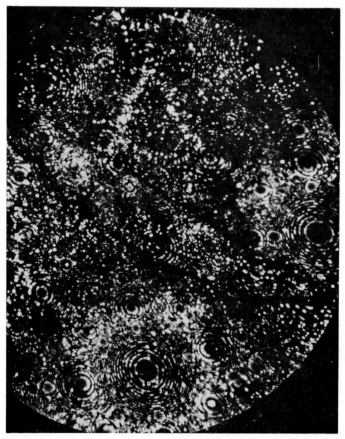

Fig. 7.13. Iron martensite: specimen of high-carbon steel, quenched from 900 °C, annealed *in situ* and imaged with helium after hydrogen-promoted field evaporation. (Courtesy E. W. Müller.)

This implies that precipitate and matrix cannot be imaged simultaneously with the same gas. It follows that, in order to study aspects of precipitation such as the structure of coherent boundaries and the loss of coherence, a system should be found where precipitate and matrix are equally "refractory". Alternatively a system might be found where the precipitate would give a helium image and the matrix a neon image, say, at the same voltage and a gas mixture would enable simultaneous viewing of the precipitate and matrix. This technique has been used with initially promising results for Fe-Mo-C by Gallot and Turner (private communication). Figures 7.12a, b and c show respectively hydrogen images of $Mo_2C$, the matrix (with the precipitate image blurred), and the result obtained with a $H_2/Ne$ gas mixture.

Schwartz (1968) has shown how to analyse field-ion data of precipitation

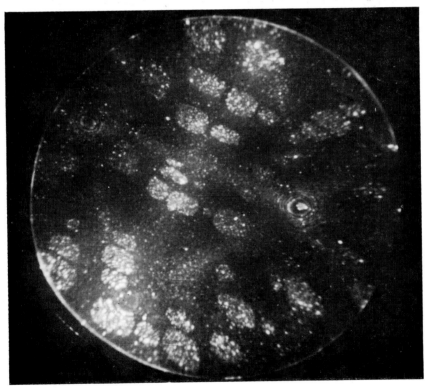

Fig. 7.14.    A nimonic type alloy showing cubes of $Ni_3Al$ precipitate.
(Courtesy R. G. Faulkner.)

in a way which permits a ready estimation of mean particle sizes. Such an averaging procedure is of course complementary to studies of individual particle morphology.

Studies of the early stages of phase transformations are naturally those where the field-ion microscope can find most useful application. Müller (1965) has obtained initially promising images of iron martensite; fig. 7.13.

The application of field-ion microscopy to the metallographic study of commercially important materials has been extended by Faulkner et al. (1968) who obtained stable images of Nimonic type alloys using neon image gas. The coherent cubes of $Ni_3Al$ appear as areas on the image of different brightness from the matrix. The sense of the brightness difference depends on how the specimen surface has been prepared. If field evaporation has been used the precipitates are dark whilst if material has been removed catastrophically by a small "flash" the precipitates are left proud of the surface and image brightly. Cubes of 50 Å side and plates 500 Å in breadth and 50 Å thick could be recognised (fig. 7.14). Meyrick (1967) has studied the dispersion of thoria particles in tungsten: a problem was that the thoria did not contribute directly to the image. Dark areas which may have been voids were attributed to the sites occupied by thoria particles which it was suggested might have been removed from the specimen by the field, e.g. in a fracture process. Occasionally image streaking was observed around the dark areas. Such image artefacts are unlikely to be caused by direct imaging of the precipitate.

A general problem in the study of alloys is the significant perturbing effect on the image of quite a small solute concentration. Ideally for precipitation studies an alloy should be found where the solute concentration in the matrix is low and there is a high volume fraction of precipitate.

# 8 | SURFACE STUDIES

## 8.1. Introduction

The first requirement of any study of surfaces or surface phenomena will usually be to obtain a surface totally free from impurities. This is widely regarded as being an impossible goal and in the past many surface studies have been made using surfaces with layers of oxide or other unwanted adsorbate. However, using the field-ion microscope and the technique of field evaporation, it is possible to obtain a surface with few impurity atoms, apart from those that would be expected from the level of impurities existing in the bulk metal.

As we have already discussed (§ 2.2) when a specimen is being imaged, say with helium at a field of about 4.5 V/Å, it is not expected that an impurity atom with an ionization field lower than that of the image gas could approach the surface. If hydrogen is used for the image gas this may not be the case, since the ionization potentials for hydrogen and nitrogen (for instance) have similar values; this means that whilst it is still possible to evaporate away pre-existing layers of oxide, the only way of preventing the subsequent approach and build-up of foreign atoms is to use an image gas with a very low concentration of impurity atoms (say $10^{-10}$ Torr of impurities).

However results from mass-spectrometry work mentioned earlier have shown that the evaporating species under certain conditions is a complex of the metal with atoms of the imaging gas or impurities. This would seem to indicate that gas atoms are in fact adsorbed on the specimen surface although they are not detected in the image.

A problem which should be borne in mind is the diffusion of impurity atoms from the specimen shank onto the imaged surface, which is always possible unless the specimen shank has been cleaned by a method such as flash-heating. One objection to flash-heating of specimens is that it tends to blunt the tip, so that the method has most application to experiments where there is no objection to working with a large (1000 Å or greater) radius specimen.

Another method of removing a stubborn surface coating is to give the specimen a rapid field-etch, sometimes in the presence of a chemisorbed gas. This method is necessary mainly for non-refractory materials. Rendulic and Knor (1968) and Rendulic (1967) found that nitrogen could attack platinum and rhenium on those planes above which the field is lower than the average value over the tip, giving rise to holes produced by enhanced field evaporation. When the adsorbate that accumulated in these holes migrated across the surface, it gave rise to enhanced field evaporation as it went. The dark $\{111\}$ areas in images of iron, and the irregular surface region around the $\{110\}$ and $\{012\}$ areas of nickel tips, have been attributed by Rendulic (1967) to the presence of chemisorbed gas on the surface. If this idea is correct the surface will consist of two regions: one part clean, imaging normally and with the expected evaporation field, and the other with a chemisorbed gas layer and a slightly lower evaporation field. The areas with the reduced evaporation field will be somewhat recessed and the image will exhibit dark or disordered patches.

A good image can be obtained by the device of introducing hydrogen, so that the entire surface becomes covered with a chemisorbed gas layer, and the difference in evaporation field between the different regions is smoothed out. Raising the field makes it possible to remove the gas layer from all over the surface, which is then left largely free from contaminants.

In addition to the advantage the field-ion microscope possesses due to the specimens having an atomically clean surface, there is a further crystallographic advantage. It is possible to study the crystallographic specificity of surface reactions, since all important planes and directions can be contained in a single field of view and direct comparison between the different regions is immediately possible.

There have been a large number of surface studies, particularly those involving vapour deposition which have used the related technique of field electron emission, and as a result there is now a considerable body of data on such phenomena as the effects of adsorbed layers on work functions, the thermal and field evaporation of adsorbed layers and the diffusion of adsorbed materials across the specimen surface. This will not be considered in any detail here since adequate reviews of the subject are available: Dyke and Dolan (1956), Gomer (1961) and Müller (1967b). It is necessary to point out though, that, particularly in vapour deposition experiments, there are strong arguments in favour of parallel field-ion/field emission studies.

## 8.2. Surface damage studies

If the imaging field is reversed in a field-ion microscope, so that the specimen is negative rather than positive, the gas ions instead of striking the screen will strike the specimen giving rise to 'cathode sputtering'. This is one reason for field emission microscopes having even more stringent vacuum requirements than field-ion microscopes.

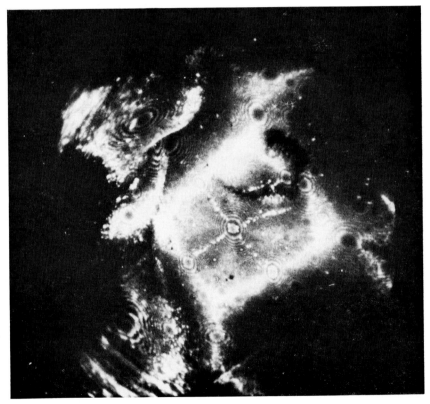

Fig. 8.1.   Surface sputtering as a result of reversing the field in the presence of the image gas. Massive damage results. Iridium imaged with helium at 78°K.

Reversal of the field whilst the imaging gas is in the microscope results in colossal damage to the specimen: point defects, dislocations and even large craters are formed. Indeed the damage is so great that it is difficult to account for it solely in terms of gas bombardment and it is probable that

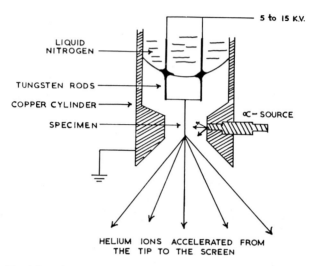

Fig. 8.2.    The experimental apparatus for *in situ* α-bombardment
used by Brandon et al. (1961).

some of the damage arises from specimen heating, both from gas impact and from the field emission current. An example of the damage produced is shown in fig. 8.1. The situation is difficult to quantify because the distance from the tip at which the incoming particles acquire their charge is not known. Müller (1960) has made a rough estimate that at a gas pressure of $2 \times 10^{-3}$ Torr some 60 ions/sec will strike each atom in the tip surface at low applied fields.

Controlled cathode sputtering has been used to remove tightly bound surface coatings from field emission specimens, and for sharpening blunt specimens, but the surface left behind is usually very ragged and unstable as a result of most of the material being removed from the side of the specimen.

Bombarding specimens with α-particles in the microscope can give rise to a more controlled form of surface damage. In studies by Müller (1959) and by Brandon, Southon and Wald (1961) images of specimens were photograph-

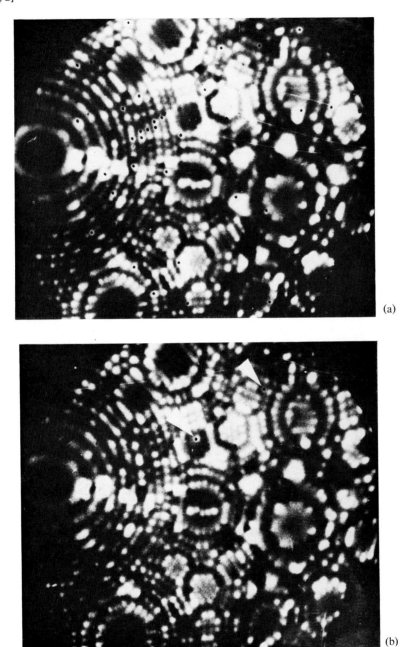

(a)

(b)

Fig. 8.3.　Result of the impact of an α-particle on a tungsten specimen in the apparatus shown in fig. 8.2: (a) before and (b) after. Atoms removed or ejected onto the surface are marked. (Courtesy M. Wald.)

ed and then an α-particle was allowed to strike the side of the specimen and the image rephotographed. The experimental apparatus used for this is shown in fig. 8.2 and typical results in fig. 8.3. In practice the α-source was arranged so that one α-particle would strike the specimen in a way which would produce damage in the field of view on average once every three hours, and photographs were taken at the rate of one every twenty minutes or less. There was then a good statistical chance that any difference between two succesive micrographs would be due to the impact of a single α-particle.

The surface acts as an integrator for the damage over what would be a much larger region in the bulk material. In the particular example shown in fig. 8.3, the total number of new vacancies seen was about 40 and 2 extra atoms appeared on the tip surface. A useful method for spotting small changes such as these in a surface is to use the colour superposition technique described by Müller (1957), which has also been used by Murr and Inman (1965), Ehrlich (1966) and Anderson (1967). The 'before' and 'after' micrographs are projected, one through a green filter and the other through a red filter, onto a common screen; unchanged atoms appear orange whilst new atoms or vacancies appear red or green.

The vacancies which appear are regarded as marking the points where a focussed collision sequence had intersected the surface of the specimen. Although such a sequence in bulk material ends in an interstitial atom, in this case it was considered that as well as the single atom which was sputtered at the end of the sequence, the second "replacement" atom would also have enough energy to be ejected for all but a few focussed collision sequences near the end of their range and it is the missing second atom which appears as a vacancy. In the presence of the imaging field, the additional energy required to eject a tungsten atom is very low – probably less than 1 eV (Southon, private communication) – but in a few cases it is expected that only one atom would have been sputtered, and in even fewer cases the focussed collision sequence would have merely pushed an atom to the surface without ejecting it.

The damage observed on the specimen surface in fig. 8.3 can be divided into two types: a compact region of intense damage and more diffuse damage over the whole surface. The distribution is plotted in fig. 8.4. If we make the assumption that all this damage is due to the intersection of focussed collision sequences with the specimen surface, some interesting conclusions can be drawn.

Nelson (1963) has concluded that for the various focussing directions in tungsten the maximum number of collisions before attenuation, $n_f$, are:

$$n_f \langle 111 \rangle = 150; \quad n_f \langle 100 \rangle = 20; \quad n_f \langle 110 \rangle = 30.$$

The radius of the specimen was 300 Å so only $\langle 111 \rangle$ sequences could have caused the widely spaced overall damage. The region of more intense damage could be due to the intersection of focussed sequences of all the main types with the surface – in which case the primary event must have occurred about 60 Å, or just less, below the specimen surface. It could also be due to there being an optimum focussing energy for $\langle 111 \rangle$ sequences which corresponds to a range which is only satisfied in this region – in other parts of the specimen the sequence has terminated before reaching the surface.

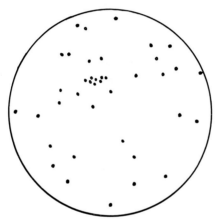

Fig. 8.4.   The distribution of damage in the α-particle impact shown in fig. 8.3.

Some possibilities can readily be ruled out. Assuming that the long-range sequences are $\langle 111 \rangle$ in direction it should be possible to pinpoint a region corresponding to the primary event from whence the sequences have propagated, provided this region does not exceed say 30 Å radius. However, this is not possible: in order to achieve a mean-free path of about 30 Å or 10 interatomic spacings the energy of the primary is required to be 30 keV. About 40 atoms visible in the image were displaced by the event (fig. 8.4). Since the solid angle viewed is about $\pi$ and assuming that only one atom in five on the surface is visible, the total number of atoms displaced is about 800. The average energy which can be allotted to each displaced atom on the basis of a 30 keV primary is then only about 40 eV and it is entirely unrealistic to postulate long-range sequences starting with this sort of energy, which is below the expected threshold energy for a single displacement. It is more realistic to assume the energy of the primary knock-on to be about 250 keV, or about $\frac{1}{2}E_{\text{max}}$ for a 5 MeV α-particle. The average energy available for each displaced atom is then about 300 eV, but the mean free-path of the primary event increases to $\approx 150$ Å or roughly half the radius of the tip. With such a large region the uncertainties of the analysis become rather extreme.

It is *possible* to interpret the damage on the basis that most of it was caused by *channelling* parallel to {110} planes, with a large concentration corresponding to the $\langle 1\bar{1}0 \rangle$ directions in the plane, one of which is roughly parallel to the tip axis. It would be expected that a channelled atom could create one or more surface vacancies close to where it is ejected from the lattice. If the damage is interpreted on this basis the size of the primary event region can be estimated from the distribution of the vacancies to be about 60 Å by 120 Å in a plane normal to the tip axis and probably about 100 Å in the third dimension.

Bombardment experiments inside the microscope are probably the most useful for gaining information about focussed collision sequences, and it would be particularly useful if such experiments were done in an ultra high vacuum microscope, so that the field could be lowered during bombardment without the specimen surface becoming contaminated. Carrying out experiments at different fields, would give a better idea of the energy of the focussed collision sequences and other events that intersect the surface.

One other method of investigating the range of irradiation events (particularly the range of the bombarding species itself) is by ion bombardment of prepared specimens either in the microscope sometimes with the image field turned off (Hudson, 1969) or outside the microscope (Buswell, 1970). The usual method is to bombard the specimen from one side, which can readily be arranged to correspond to a particular crystallographic direction if required. The approximate depth of penetration of the bombarding species can be measured by field evaporating through the bombarded region and recording the distribution of damage across the specimen.

In one experiment (Hudson et al., 1968) iridium cooled by liquid nitrogen has been irradiated with 100 keV Xe$^+$ ions, and fig. 8.5 is a helium ion micrograph of an iridium specimen irradiated to a dose of $10^{11}$ ions·cm$^{-2}$. The bombarding species (Xe$^+$) was chosen to have similar effects to Ir$^+$ ions, high doses of which cannot readily be obtained. A dose of $10^{11}$ ions· cm$^{-2}$ will not give significant overlap of displacement cascades, since only one or two primary events should occur in the volume of the specimen that is examined.

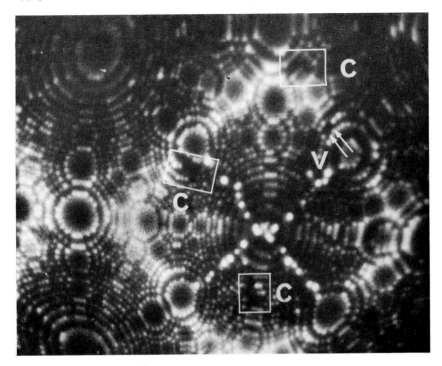

Fig. 8.5. Iridium specimen irradiated at 78°K with 100keV Xe⁺ ions to a dose of $10^{11}$ ions·cm⁻². The positions of clusters (c) and vacancies (v) are indicated. (Courtesy J. Hudson.)

The damage observed in this type of experiment takes the form of seemingly isolated single vacancies. However, it has been shown that plotting the vacancies on a 3-dimensional model enables the identification of dispersed clusters in the damage, that is distinct regions that have a high concentration of vacancies. In the work of Hudson et al. (1968) the results of this analysis are somewhat anomalous because the damage appeared to be spread right across the specimen, indicating an ion penetration of about 800 Å (the diameter of the specimen). This is nearly an order of magnitude greater than the calculated mean range, but the specimen in this case was bombarded close to a ⟨100⟩ direction, a prominent channelling direction in face centred cubic metals. It seems likely that the damage has been spread across the specimen as a result of the influence of channelling and focussed collision sequences, although some arguments against this explanation were put forward by Hudson et al. (1968).

## 8.3. Oxidation

Under certain conditions it is possible to image oxygen adsorbed on the surface of a field-ion specimen. The initial oxidation can be achieved by exposing the specimen to a small partial pressure of oxygen for a short time, and subsequent imaging of the adsorbate is possible at relatively low fields.

Bassett (1966) has used field emission techniques to establish the conditions under which oxygen adsorbed on tungsten may be imaged. He found that the oxygen desorption was greatly enhanced by the presence of helium or neon imaging gas, and that it was removed extensively, if not completely as a complex with the tungsten. With liquid nitrogen specimen coolant, the removal of the adsorbed layer was not completed in vacuum until the evaporation field for tungsten was reached, whilst in the presence of a few millitorr pressure of helium or neon, complete desorption had taken place by 85% and 77% respectively of the evaporation field. The imaging voltage for neon is lower than that for helium due to its lower ionization potential whilst the tungsten evaporation fields in the presence of either gas are very similar. Bassett concluded that it should be possible to image oxygen adsorbed on tungsten using neon as the imaging gas and at as low a temperature as possible. Hydrogen was not investigated as a possible imaging gas.

Cranstoun and Anderson (1966), however, have shown that it is possible to obtain significant results using hydrogen as the image gas for the adsorption of oxygen on tungsten. Using fields of about 2 V/Å, i.e. below hydrogen B.I.V., they found that it is possible to observe scintillating spots that are associated with the presence of oxygen, particularly on the zones between $\{110\}$ and $\{12\bar{1}\}$. They found that the number of spots could be varied by changing the oxygen adsorption conditions and was directly proportional to the pressure of oxygen, other parameters being kept constant. The scintillation rate was proportional to the pressure of the hydrogen image gas.

Using higher pressures of oxygen it is possible to see an effect that could correspond to field induced migration of a second layer of oxygen from the shank. An oxygen pressure of $10^{-6}$ Torr and a field of 2 V/Å can induce field corrosion which results in the formation of craters in $\{111\}$ regions with channels radiating towards $\{211\}$ and $\{110\}$. This has been interpreted (Cranstoun and Anderson, 1968) in terms of surface rearrangement as a result of the specific crystallographic nature of diffusion paths for gases over the surface.

Fortes and Ralph (1968b) have shown that it is possible to get thick films of oxide if desired, by exposing iridium to oxygen at atmospheric pressure above a critical temperature (540 °C); a polycrystalline film of

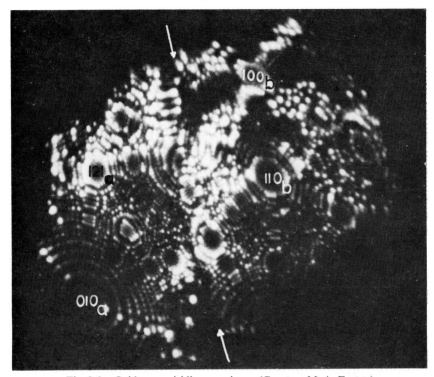

Fig. 8.6.    Oxide on an iridium specimen. (Courtesy M. A. Fortes.)

iridium oxide is developed. The method would appear to have a potential application in measuring the kinetics of oxide growth, but the difficulty in using field evaporation to measure the thickness of an overgrowth is that it is only possible to measure thickness accurately once something like a regular field evaporated endform has been established. The oxidised endforms were rather jagged. The oxide would begin to image as just a few spots and then gradually spread until the entire tip was imaged as shown in fig. 8.6. The films are invariably polycrystalline, and this might provide a method for studying boundaries in ceramics provided that the crystallites were large enough to be indexed.

## 8.4. Other gas-surface reactions

### 8.4.1. CARBURISING

The metal carbides provide a very interesting and commercially important group of compounds. In addition carbon in interstitial solution has considerable importance, especially in the case of steels, (Honeycombe, 1968). Carbides in general are hard, brittle materials that are difficult to prepare as field-ion specimens; furthermore in many cases, as with nitrides and oxides, they can exist over a wide range of compositions. For example titanium carbide TiC, has the general composition $TiC_x$ where $x$ lies between one and about 0.5 and there is thought to be a concentration of $(1-x)$ vacancies on the carbon sub-lattice.

It is possible to use the field-ion microscope to examine specimens prepared either from bulk carbides or by carburising a metal specimen in the microscope. The images are markedly less regular than those obtained from pure metals or ordered alloys; fig. 8.7 is a micrograph of a $TiC_{0.8}$ specimen, prepared from bulk carbide, imaged with helium ions at 27°K. The axis of such carbide specimens is invariably an ⟨001⟩ direction since the specimen is polished from a fine bar (say 1 mm square) produced by cleaving on {100} planes. The tetrad symmetry is obvious in fig. 8.7 where two poles are repeated four times at intervals of ninety degrees about the central axis. Assuming a stereographic projection the poles are indexed as {111} and {011}. The regularity of the images from titanium carbide does not improve as stoichiometry is approached, and micrographs from $TiC_{0.97}$ are no better in quality than fig. 8.7.

Non-stoichiometric zirconium carbide, $ZrC_{0.875}$, has given very similar images to the example shown of titanium carbide, when imaged with hydrogen at 78°K. Tantalum carbide gives somewhat better images (Meakin, 1968), see fig. 8.8, though still poor by comparison with those from pure metals and dilute alloys. It appears that 3% of vacancies on the carbon sub-lattice are sufficient to give an irregular field-ion image, but that greater concentrations of vacancies do not noticeably change the character of the field-ion images.

Tungsten carbide gives very similar results (French and Richman, 1968 and Cinquana and Meakin, 1967). Carbide specimens frequently show streak contrast rather than the more normal spot contrast. In some cases the streaks arise from electropolishing deposits and can be removed either by field evaporation, or more usually by cleaning with the corrosive part of the electropolishing solution which in many cases is hydrofluoric acid. Other image streaking (see ch. 3) arises in specimens with a highly elliptical cross-section, which are easily developed when specimens are prepared by a process that starts with cleavage on two planes at right angles. Figure 8.9 is a scanning electron micrograph of a specimen of titanium carbide which initially had a rectangular cross-section.

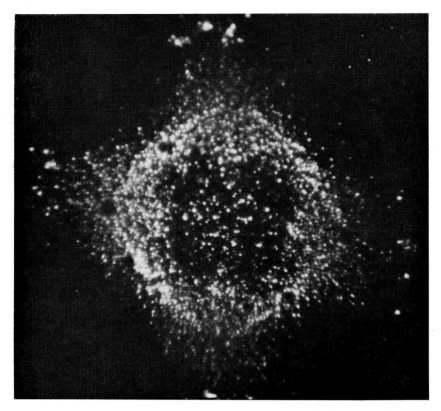

Fig. 8.7. $TiC_{0.8}$ imaged with helium at $27\,°K$. TiC has the rocksalt structure (cubic F) and because 4-fold symmetry is apparent, the axis of the specimen must be $\langle 100 \rangle$.

Cementite, $Fe_3C$, and other carbides present as secondary phases in steels can be successfully imaged using hydrogen as the image gas at $78\,°K$, as discussed in § 7.3.

8.4.2. NITRIDING

Metal nitrides give very similar images to the carbides, (Tsong and Müller, 1966b). It might be of interest to make an *in situ* study of the surface nitriding of steels (an industrial process used to provide a hard surface coating), by nitriding specimens in the microscope. It should be possible to investigate how penetration varies with time and orientation, also whether nitrides are formed and if so how they interact with dislocations. Examining specimens in both the electron microscope and the field-ion microscope might

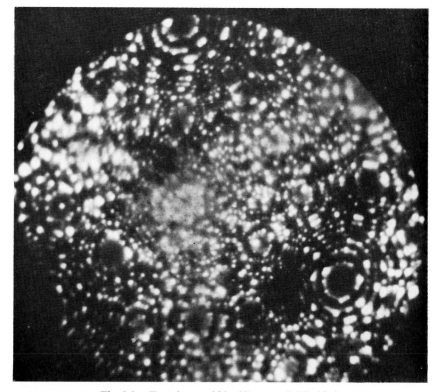

Fig. 8.8.    Tantalum carbide. (Courtesy J. Meakin.)

be useful in order to correlate measurements on the thickness of the nitrided layer, and possibly to determine by means of a diffraction pattern if there were any nitride present.

8.4.3. HYDROGEN

The interaction of hydrogen with the specimen surface and its use to promote field evaporation has been considered in § 2.4. Carbon monoxide and water vapour (Mulson and Müller, 1959 and 1963) can have similar corrosive effects.

8.5. Metal on metal

Müller first demonstrated that it is possible to image tungsten adatoms vapour deposited onto a tungsten tip in 1957. The study of thin evaporated films of metals and of epitaxy calls for more rigorous experimental condi-

Fig. 8.9.   Scanning electron micrograph of an asymmetric titanium carbide specimen.

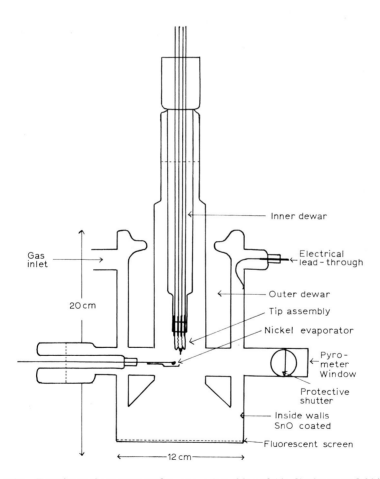

Fig. 8.10.   Experimental arrangement for vapour deposition of thin-film layers on field-ion specimens. (Courtesy G. D. W. Smith.)

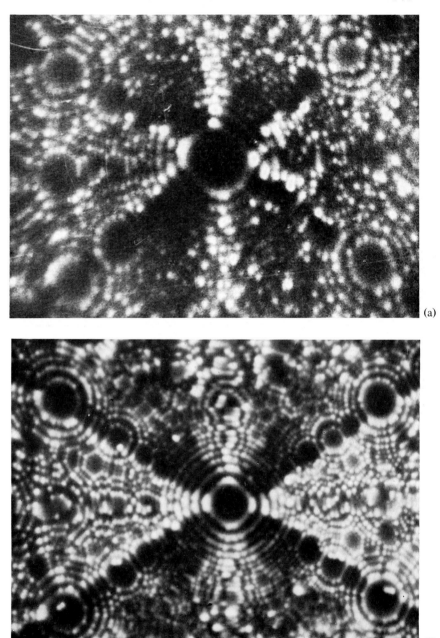

(a)

(b)

Fig. 8.11.   Thin layer of molybdenum evaporated onto a tungsten substrate; the characteristic bright triangular contrast of molybdenum can be detected. (Courtesy L. Gillott.)
Compare (b) a molybdenum specimen in approximately the same orientation.

tions than some field-ion experiments, but provides a very useful method of studying binding energies. Plummer and Rhodin (1968) have studied the binding of transition elements (Hf, Ta, W, Re, Os, Ir, Pt, Au, Mo, Rh, Pd) on the low index plane facets of tungsten. Nickel on tungsten has been studied by Jones (1966) and by G. D. W. Smith (1968) and Gillott and Southon (1967 and 1968) have examined evaporated layers of molybdenum, platinum, gold and tungsten on tungsten. Whitmell (1968) has deposited iridium on tungsten at higher temperatures in an attempt to form alloys.

The experimental conditions are fairly standard. Ultra high vacuum ($10^{-10}$ Torr or better) is mandatory and this is usually achieved either by well-trapped mercury diffusion pumps or using ion-pumps. A small evaporating source is located close to the specimen (see fig. 8.10), and the whole microscope must be capable of being baked at temperatures up to about 300 °C.

It is clear from the work that has been published that some metals show their characteristic field-ion pattern even if they are only 2 or 3 monolayers thick. Figure 8.11a shows a thin layer of molybdenum evaporated onto a tungsten substrate: the characteristic bright triangular contrast of molybdenum can be detected. The thickness of the layer has to be measured by subsequent field evaporation, and the strength of the field required for evaporation provides a measure of the binding energy. Table 8.1 shows the binding energies measured for various metals on tungsten, found by measuring the desorption field for the adatom compared with the desorption field for a tungsten adatom at the same crystallographic position.

The experimental method is to prepare a tungsten tip by field evaporation

TABLE 8.1
Binding energies in eV (after Plummer, 1967)

| Adatom | {110} | {100} | {112} | {111} |
|--------|-------|-------|-------|-------|
| Hf | — | 7.5 | — | — |
| Ta | 8.4 | 7.8 | 7.1 | 7.0 |
| W | 8.2 | 8.0 | 6.9 | 6.7 |
| Re | 10.2 | 9.3 | 8.3 | 8.0 |
| Os | 8.3 | 8.0 | 6.9 | 6.7 |
| Ir | 6.0 | 6.9 | 4.1 | 4.3 |
| Pt | 5.6 | 5.1 | 2.4 | 2.8 |
| Au | 3.8 | 3.3 | — | — |
| Mo | 9.5 | 10.5 | 8.3 | 8.7 |
| Rh | 7.0 | 6.9 | 3.5 | — |
| Pd | 5.8 | 4.2 | — | — |

at 21 °K, and then turn off the high voltage and heat a fine tungsten wire close to the specimen by resistance heating to 2800–3000 °K at which temperature the vapour pressure is $10^{-8}$–$10^{-7}$ Torr. The deposited adatoms of tungsten are removed by pulsing the image field above the best image voltage until all the adatoms have been removed, noting the voltage that is necessary to remove the atoms from each facet of the tungsten surface. The procedure is then repeated for a second metal from a different evaporator, and the desorption voltages noted.

Melmed (1966) has grown single crystals and polycrystals of Ag, Cu, Pb, Y and Fe on tungsten point substrates of larger radius and published some thickness measurements of epitaxial copper films. Montagu-Pollock et al. (1968) have reported preliminary observations of iron deposited on iridium and gold deposited on tungsten.

Defects apparent in the thin films of iridium on molybdenum published by Whitmell (1968) show that a vapour deposited layer may contain some very high energy defects e.g. iridium normally fcc (ABCABC stacking) appears to have been deposited in an ABAB sequence.

## 8.6. Diffusion studies

Any attempt to investigate surface adsorption must usually also involve surface diffusion, because an atom will not in general be adsorbed in the position at which it first arrives on the surface but in some other lower energy site. Ehrlich and Hudda (1960 and 1962), therefore, when studying the adsorption of nitrogen on tungsten specifically extended the study to include migration of the adsorbed species.

Adatom self-diffusion of tungsten on tungsten has been investigated by Ehrlich and Hudda (1966) and by Gillott (1968) and Parsley (1969). The experimental method is to deposit adatoms from an emitter in the usual way, with the tip at a low temperature, and to photograph the surface by introducing an image gas and applying high voltage to the specimen. Diffusion of the adsorbed atoms over the surface is allowed to take place by raising the temperature (§ 2.3) to the required value for a standard time (say 5 min). The specimen is not exposed to the high voltage or the imaging gas during the diffusion operation, but these are reintroduced afterwards to enable the surface to be imaged again. It is usual to make a series of perhaps five diffusion cycles and then to repeat the experiment at higher diffusion temperatures. Ehrlich and Hudda analysed their data in terms of the mean square displacement of an atom from its original position after the surface

TABLE 8.2
Adatom surface diffusion parameters for tungsten, the first two lines refer to self-diffusion, whilst the work of Bassett and Parsley is on rhenium adsorbed on tungsten

| Plane: | Activation enthalpy (kcal/g·atom) | | | Activation entropy (e.u.) | | | Pre-exponential factor (cm²/sec) | | |
|---|---|---|---|---|---|---|---|---|---|
| | {110} | {211} | {321} | {110} | {211} | {321} | {110} | {211} | {321} |
| Ehrlich and Hudda (1966) | 22 | 13 | 20 | $4 \pm 4$ | $-20 \pm 10$ | $-3 \pm 3$ | $3 \times 10^{-2}$ | $2 \times 10^{-2}$ | $1 \times 10^{-2}$ |
| Gillott (1968) | $16.6 \pm 1.6$ | $8.8 \pm 0.8$ | — | $-16 \pm 10$ | $-22 \pm 10$ | — | — | — | — |
| Bassett and Parsley (1969) | 23.9 | 20.3 | 20.4 | — | — | — | $3 \times 10^{-2}$ | $11 \times 10^{-3}$ | $5 \times 10^{-4}$ |

had been heated for a fixed time at a given temperature; Gillott used an alternative analysis based upon an observation of the fraction of atoms moving under these conditions corrected to allow for multiple jumps. Table 8.2 compares *inter alia* the measured values for activation energy and entropy on the $\{110\}$ and $\{211\}$ planes; it will be seen that there is reasonably good agreement between the two sets of values. Ehrlich and Hudda also reported that the mobility at room temperature increased in the order:

$$(211) > (321) \approx (110) > (310) \approx (111).$$

Brenner (1965) has used the field-ion microscope to study the surface diffusion of iridium both in ultra-high vacuum and in the presence of gaseous impurities. He used field evaporated iridium tips of radius 400–800 Å, which were heated for various times and at temperatures up to 1000 °C in the absence of an applied field. The specimens were heated by the standard method of resistively heating the wire loop onto which the specimen was spot welded, and measuring the temperature by following the change in resistance. The effects of surface migration were studied by subsequent field evaporation. Brenner found that appreciable surface rearrangement took place above about a quarter of the melting temperature (450 °C). The re-arrangement took the form of a dissolution of the high index plane facets, followed at higher temperatures by atom migration from the edges of low index planes (which requires atom migration over larger distances). Figure 8.12 shows a typical surface, before restoration to its more usual structure by field evaporation. Oxygen was found to alter the surface topography after rearrangement by increasing the surface mobility of some iridium atoms, but the presence of 0.1 Torr of hydrogen, nitrogen or carbon monoxide had little effect.

A study of the thermal rearrangement of tungsten surfaces has been made by Bassett (1965) using similar methods. He observed extensive surface rearrangement above 600 °C, with minor rearrangements taking place at lower temperatures down to about 25 °C. One great advantage of the field-ion microscope in surface diffusion studies, is that it is possible to obtain values for the activation energy for surface self-diffusion *in particular crystallographic directions*. Bassett, for instance, found that at 500 °K in tungsten, atoms were displaced along the edges of $\{011\}$ terraces from one kink site of $\{111\}$ type to the next, with an activation energy of 1.8 eV, while displacement from the $\{001\}$ type kinks had an activation energy of 2.1 eV and took place at a slightly higher temperature. Similar displacements along the edges of $\{111\}$ and $\{001\}$ terraces had activation energies of 2.2 and 2.3 eV

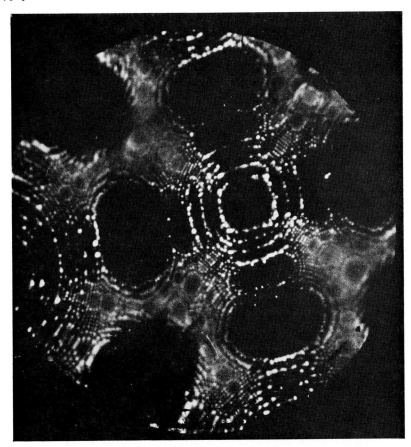

Fig. 8.12.   Surface of iridium specimen after high temperature rearrangement.
(Courtesy S. S. Brenner.)

respectively, and took place at 600 to 700 °K. Figure 8.13 shows one of the
effects of heating the tungsten surface to higher temperatures where massive
surface rearrangements can take place. Long unbroken rows of close-packed
atoms tend to be produced, often destroying the regular ring pattern of the
normal field evaporated endform, and apparently producing some layer
edges with steps more than one atom layer high.

Bassett and Parsley (1969) who have looked at the diffusion of adatoms
of other elements (rhenium and iridium) on tungsten, also made the interesting observation that the diffusion coefficient of rhenium on tungsten is not
significantly altered when diffusion occurs in an applied field that corresponds
to a voltage about 20% of the best image voltage for the tip.

## 8.7. Electrodeposited layers

The field-ion microscope is a powerful tool for studying electrodeposits because field evaporation makes it possible to dismember the deposited layer and study its atomic perfection, and also to investigate the nature of

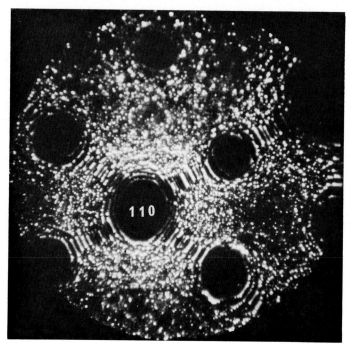

Fig. 8.13.  Surface of tungsten specimen after rearrangement above 700 °C.
(Courtesy D. Bassett.)

the interface between substrate and deposit. Rendulic and Müller (1966a and 1967a) have studied electrodeposits of platinum and rhodium on tungsten and on iridium field-ion tips. Standard electroplating solutions can be used, for instance for platinum a solution (Rendulic and Müller, 1967b) of sulfato-dinitro-platinous acid $H_2Pt(NO_2)_2SO_4$ with a concentration of 5 g/litre, at a temperature of 50 °C and a $p$H of 1. With this solution Rendulic and Müller obtained best results by using 60 cps AC at 1.2 V and 500 $\mu$A for 30–60 sec.

Electrodeposits tend to be highly polycrystalline with a wide range of crystal sizes that have been variously reported as 50–500 Å or 700 Å average, and usually have either $\langle 111 \rangle$ or $\langle 113 \rangle$ normal to the surface in the case of platinum deposits.

The addition of various agents to the electroplating bath can markedly affect the nature of the electrodeposit; Coumarin and vitamin $B_{12}$ nearly halve the average crystallite size and regularise the orientation of the crystallites of electrodeposit. It should be possible to study the effects of various surface "brighteners" that are sometimes added to electroplating baths, the precise function of which is not at present very well understood.

## 8.8. Macro-molecules

One of the reasons for interest in electrodeposition is to see whether it is possible to incorporate large biological molecules in the deposit, and then to investigate the shape of these by a type of replica technique. Although it may not be possible to obtain any sort of image from a molecule itself, it should be possible to measure accurately the shape and dimensions of the hole which has been occupied by the molecule. Rendulic and Müller (1966) have obtained helium ion images thought to come from coenzyme I (DPN, diphosphopyridene nucleotide) and phthalocyanine radicals in a platinum matrix. Bell, Cooper and Adams (1966) using a similar technique have obtained structureless spots approximately 10 Å in diameter using hydrogen as the imaging gas, and thought to have come from single phthalocyanine groups. They also carried out field stability tests by examining the field electron emission pattern in vacuo, and concluded that the phthalocyanine complex was stable up to the following fields:

1.8–2.3 V/Å in vacuo

1.2–1.5 V/Å in the presence of $10^{-3}$ Torr of hydrogen

1.5–1.8 V/Å in the presence of $2 \times 10^{-3}$ Torr of helium.

An alternative approach has been used by Nadakivikaren and Hutchinson (1966) who have "vapour" deposited phthalocyanine radicals and tungsten atoms onto a clean tungsten substrate and attempted to image the molecules in this way.

Melmed (1966) has imaged copper phthalocyanine and nylon in parallel field-ion and field-emission experiments.

# 9 | STUDIES WITH ALLOYS

## 9.1. Introduction

Application of field-ion microscopy to alloys is potentially very fruitful since there remain fundamental problems in alloy behaviour which may require atomic resolution for their complete explanation. Interpretation of alloy images requires extension of the elemental theories of field evaporation and the imaging process. Since the field evaporation process is used to complete specimen preparation prior to imaging it is appropriate to consider first the field evaporation behaviour of an alloy.

## 9.2. Contrast from alloys

In general a solute atom in an alloy may be evaporated at a field lower than, the same as, or higher than the matrix. A calculation by Brandon (1966f) to estimate the binding energy of a solute atom in terms of known thermo-dynamic quantities has been extended and refined by Southworth (1968) and Southworth and Ralph (1968). The binding energy can be substituted in eq. (1.9) in order to estimate the evaporation field, $F_E$, for the solute. Both the field acting on and the binding energy of an atom depend on the envi-ronment of the site occupied. In particular for preferential evaporation of solute to occur from within a plane ring the decrease in $F_E$ for the solute, $\Delta F$, must be greater than the usual decrease in field that occurs above the atoms in the centre of a low-index plane, $\Delta F^*$. I.e. if $\Delta F - \Delta F^* > 0$, then preferential field evaporation is expected.

Preferential retention can be treated similarly, but there may be an additional contribution to the binding energy due to a polarization bonding contribution as for the atoms constituting zone decoration (see § 3.2.4). Cranstoun (private communication) has some evidence that, in iridium, surface re-arrangements may be induced in the vicinity of interstitially dissolved oxygen. He proposed that a matrix atom is stabilised in a protruding site by polarization bonding above the solute, thus accounting for the bright contrast associated with non-metallic interstitials, such as oxygen in iridium, which was described in ch. 4.

Ralph and Brandon (1963) suggested that the field above a solute atom could be modified by charge transfer to the matrix. Intuitively one expects the ionization probability to be greater above a centre with enhanced positive charge. The corresponding image spot intensity will also be greater. However an increased positive charge implies a locally higher field with a correspondingly greater tendency towards preferential field evaporation. For instance gold, palladium, nickel and tungsten are all positive with respect to a platinum matrix and are expected to image brightly *if* they are not first removed from the specimen surface by preferential field evaporation.

Dubroff (1967) studied a series of dilute alloys of platinum; vacant site contrast was observed for alloys with gold, palladium and nickel and bright spot contrast with tungsten. The calculations of Southworth and Ralph (1968) suggested that the solute would be preferentially evaporated from these alloys except tungsten which could be preferentially retained.

Some reasons for variations in image intensity from the different species in an alloy have been proposed by Ralph and Brandon (1963) and Tsong and Müller (1968). It is suggested that the image spot intensity could be related to the sense of the charge exchange between the two species. Some measure of the attractive power of electrons is required and Southworth and Ralph (1968) have pointed out that the reported values of the work function of solute and solvent in many alloy systems appear to provide a useful guide to the sense of the charge shift.

Tsong and Müller (1968) attempted to explain the contrast in alloy images exclusively on the basis of image intensity variations. Since field evaporation *precedes* field-ionization it is clear that when preferential field evaporation occurs, consideration of relative brightness may be irrelevant. In particular, Southworth and Ralph's (1968) calculations showed that the invisibility of cobalt in alloys with platinum was consistent with preferential field evaporation rather than charge exchange.

The brightness of an image spot depends *inter alia* on the relative prominence of the site occupied by the source atom. When preferential retention occurs, as in platinum-tungsten and tungsten-rhenium, the solute is stabilised in a protruding low co-ordination number site and accordingly might be expected to give a bright image spot, providing any opposing charge exchange has less effect. Computer simulations (Moore, 1967, and Moore and Ranganathan, 1967) provide a corroborative illustration for the consequences of the effects analysed by Southworth and Ralph (1968). Striking computer simulated micrographs were produced showing the disruptive effect of selective evaporation on image regularity. This has also been observed experimentally (Ralph and Brandon, 1963 and Caspary and Krautz, 1965).

### 9.3. Studies of solid solution alloys

The regularity of the field-ion image decreases as the solute content rises (Brandon and Ralph, 1963). The effect is illustrated in fig. 9.1 for the series of tungsten-rhenium alloys studied by Brandon and Ralph; with 5% rhenium the image is only slightly disrupted, but at higher concentrations only low index planes are developed recognisably Similar effects have since been

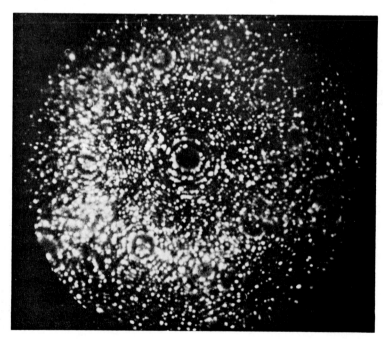

(a)

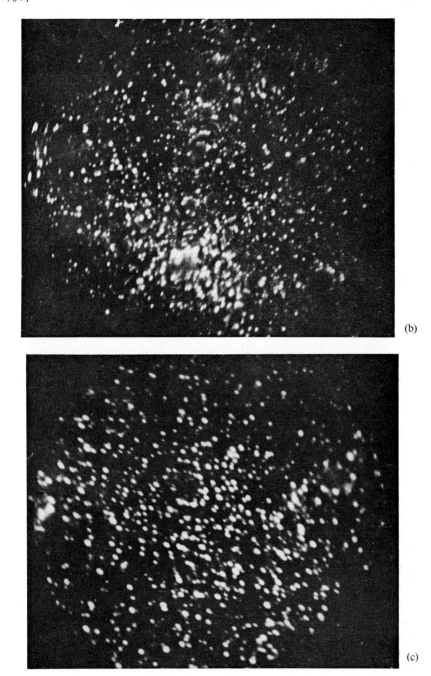

(b)

(c)

Fig. 9.1.   Series of micrographs showing the effects of increasing solute content: (a) tungsten – 5% rhenium (b) tungsten – 26% rhenium (c) tungsten – 34% rhenium. (All at 78°K, helium image gas.) (Courtesy B. Ralph.)

found for a variety of alloy systems. Although the regularity of the image has deteriorated, crystallographic analysis can still be done; for instance in fig. 9.1a the central pole can be identified as {110} but the absence of diad symmetry overall, indicates the presence of a grain boundary.

Identification of the species is a crucial problem in studies of solid solution alloys. An attempt at identification can be made from the imaging behaviour but difficulties may arise such as confusion between lattice vacancies and those induced by preferential evaporation of a second species. Similarly, bright spots may arise from a number of causes. In any case the site occupied by a polarization bonded atom may not be a lattice site making it difficult to decide exactly where a solute was in the bulk material.

An interesting correlation between the alloy content of a material and the magnitude of a discontinuity in a plot of the number of image spots from one crystallographic region of a specimen against applied voltage has been reported by Ralph (1966). Potentially this could provide a powerful technique for micro-analysis. It appears however that the method is neither generally applicable nor reproducible (Brenner, 1967; Whitmell, 1965). Elvin (1967) showed that there was a relationship between concentration and the proportion of preferentially retained rhenium atoms in tungsten-rhenium alloys but not a one to one correspondence. In a detailed study of several systems Dubroff and Machlin (1968) have found a correlation between the number of extra bright spots or apparent vacant sites and solute concentration. The atom probe field-ion microscope (§ 2.7.4), promises to be a powerful tool in alloy studies since it provides the means for atomic analysis to complement the atomic resolution of the field-ion microscope.

Segregation and clustering can be studied in atomic detail. Such investigations might, for instance, clarify the behaviour of embrittling solutes in bcc materials. Short range ordering and clustering, the inverse phenomenon, have been directly revealed in platinum-2 at% nickel and platinum-2 at% gold respectively (Gold and Machlin, 1968). The positions of solute atoms were determined and these data were analysed to determine short range order coefficients, cluster distribution functions and morphology. Nearest neighbour clusters were found in the gold-platinum solid solution; ordering was found in nickel-platinum. The clusters found in platinum-2 at% gold were planar and predominantly on {111}. Interaction energies calculated from the field-ion data were found to be in good agreement with those derived from thermochemical data. Images are highly regular for solid solutions of platinum and iridium (Müller, 1962). For this system $\Delta F \simeq 0$, (§ 9.2).

## 9.4. Order-disorder

The contribution which field-ion microscopy has made and could make to the understanding of the properties of ordering alloys will now be considered. Müller (1962) demonstrated that regular images could be obtained from fully ordered cobalt-platinum; Ralph and Brandon (1964) showed that in cobalt-platinum the permanent magnetic state corresponded to partial order indicated by incomplete image regularity (fig. 9.2). The domain structure is of great importance because it controls the properties of order-disorder alloys. Rotational and translational domain boundaries can be identified and characterised in field-ion images by noting orientation changes and ring mismatching. Translational anti-phase domain boundaries

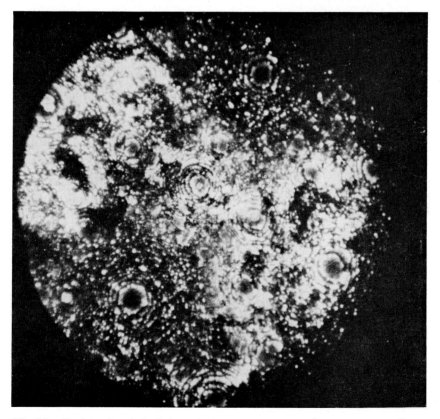

Fig. 9.2.    Equiatomic cobalt-platinum in the permanent magnetic state corresponding to partial order. (78 °K, helium image gas.) (Courtesy B. Ralph.)

give contrast equivalent to a stacking fault and their images obey the same rules. A distinction could be made between anti-phase domain boundaries having different $q$ values by noting the relative ledge radii (§ 5.6). Rotational domain boundaries (such as that shown in fig. 9.3) may be analysed in the same way as high angle grain boundaries (§ 1.8.5 and § 6.2).

Southworth (1968) has made a distinction between the mechanisms of ordering at 660 °C and 500 °C in platinum-cobalt by a combined field-ion microscope and X-ray study. At the higher temperature rapid homogeneous ordering occurred, rather than nucleation and growth, whilst at the lower

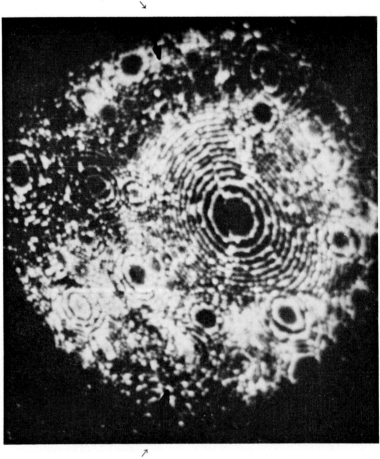

Fig. 9.3.   Rotational ordered domain boundary in equiatomic cobalt-platinum (78 °K, helium image gas). (Courtesy H. N. Southworth.)

temperature it was found that large ordered domains (with a diffuse inter-face) had swept through the short range ordered matrix.

In ordered alloys it is sometimes relatively easy to distinguish the two species and it has been shown (Southworth and Ralph, 1966; Tsong and Müller, 1967a and 1967b) that misplaced atoms give "wrong" contrast, e.g., in cobalt-platinum a misplaced platinum atom gives a bright spot in an unexpected place and a misplaced cobalt atom gives no spot where one would otherwise be expected.

Nickel-molybdenum is another order-disorder system which has been studied by field-ion microscopy (Newman and Hren, 1967; Lefevre et al. 1968). Langdon and Dorn (1968) have suggested complex dislocation disso-ciations in ordered copper-gold, $Cu_3Au$, to account for its high Peierl's stress. As in the case of bcc metals, the stacking fault widths predicted are very small. Nevertheless it is possible that during imaging the width might be increased in the same way as the widths of faults in tungsten and iron and be detectable by field-ion microscopy. The theory of ch. 5 would provide a suitable framework for analysis. Copper-gold alloys have already been

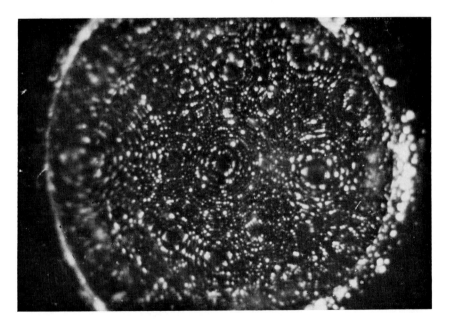

Fig. 9.4.   Ordered copper-gold, CuAuII; image recorded using an image converter (63°K, neon image gas). (Courtesy P. J. Turner.)

shown (Turner, 1967) to yield good field-ion images (e.g. fig. 9.4) using neon as image gas and a converter system; some details of the periodic antiphase domain boundary structure in Cu-AuII have been analysed by Southworth (1968).

Finally we consider the implications of the invisibility of one species on the relative prominence rule of Drechsler and Wolf (1958). In equatomic cobalt-platinum (fct) for instance, the most prominent plane is (001) not $\{111\}$ since, considered in terms only of the platinum atoms which contribute to the image (the cobalt is not imaged) it has the biggest step height. A similar effect has been found (Newman and Hren, 1967) in ordered nickel-molybdenum, $Ni_4Mo$, (fig. 9.5). To fit the observations the relative prominence rule must be re-written in terms of the interplanar spacing for the imaged species *only*. This is consistent with computer simulations (Moore and Ran-

Fig. 9.5.   Ordered nickel-molybdenum, $Ni_4Mo$, photographed using the fibre-optic screen technique. Domain boundaries are visible. (Courtesy J. Hren.)

ganathan, 1967) where geometrical justification is also provided for the relatively small number of planes developing on field-ion tips of ordered alloys and pure metal specimens of the same radius.

APPENDIX **1**

# THE APPLICATION OF COMPUTER TECHNIQUES TO FIELD-ION MICROSCOPY

## A1.1. Introduction

As we have seen the interpretation of field-ion images is a problem of some complexity for at least two reasons. Firstly there is a large amount of information on a single micrograph, relating to some $10^4$ image spots; and secondly the imaging criteria especially for a defect image are not entirely obvious.

Both problems are potentially more amenable to study by computer techniques than by "long-hand" methods owing to the need for frequent repetition of similar operations.

## A1.2. Computer simulation of perfect field-ion images

Moore (1962) showed the feasibility of computer simulation of field-ion patterns. His method is to instruct the computer to plot the projected positions of those atoms of which the sites lie within a thin hemispherical shell which is imagined to intersect the lattice. The semi-quantitative agreement with a real field-ion micrograph is illustrated in fig. A1.1. Instead of using a pen plotter for producing the image, it is possible to use a computer with direct visual display and photograph the "image" (an example is shown in fig. A1.2). Sanwald and Hren (1967) have shown that an image spot to atom correspondence may be obtained by considering co-ordination number as an additional imaging criterion together with local radius changes; fig. A1.3 illustrates the good correspondence achieved using this 'neighbour' model for the (931) region of platinum. However this model is difficult to apply to the whole area of a micrograph and has not been used for defect images. Consideration by computer methods has shown that atoms in kink sites are those which contribute to the field-ion image (Brandon and Perry, 1967a).

Moore (1967) and Moore and Ranganathan (1967) have extended the shell model to include alloy images; invisibility of one species (selective ionization or evaporation) is taken into account by using different shell thicknesses for matrix and solute. The simulations show the increase of dis-

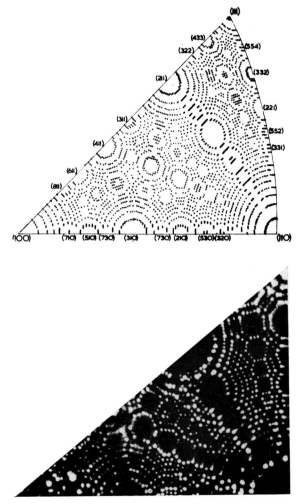

Fig. A1.1.   Comparison of actual and simulated field-ion images: shell model.
(Courtesy A. J. W. Moore.)

Fig. A1.2. A simulated field-ion image photographed directly from the visual display screen of a PDP7 computer (modified shell program). (Courtesy M. Taylor.)

order in the image with increasing solute concentration and degree of selective field evaporation. In particular the asymmetry of image irregularity about the equiatomic composition in tungsten-molybdenum alloys is shown to be consistent with the greater randomising effect of preferential retention of tungsten in molybdenum rich alloys than preferential evaporation of molybdenum from tungsten rich alloys as discovered by Caspary and Krautz (1965).

A limitation to the usefulness of computer models at present is the difficulty in relating the parameters used to obtain realistic computer 'images' to the physical processes of imaging.

### A1.3. Computer simulation of defect images

The dislocation contrast theory described in ch. 5 relies on the same

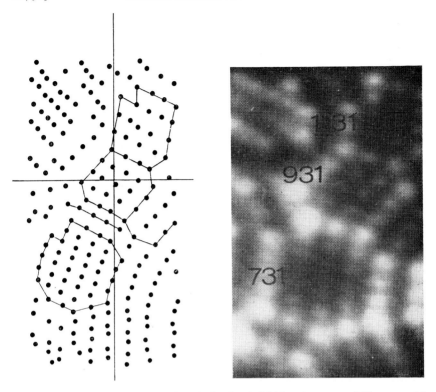

Fig. A1.3.   Comparison of actual and simulated field-ion images: neighbour model.
(Courtesy R. Sanwald and J. Hren.)

imaging criterion as computer simulations. Good agreement is found between
computer simulations by Brandon and Perry (1967b, 1968), Sanwald and
Hren (1968), Ranganathan (1969), Ranganathan et al. (1966) and the
intuitive approach of ch. 5. For instance, figs. A1.4a and A1.4b taken from
the work of Brandon and Perry are simulated examples of perfect and
dissociated dislocation contrast respectively. A difficulty in the application of
computer simulation is the presence of extra image spots as an artefact of the
technique. Neither approach takes account of secondary effects (such as
preferential evaporation, distortion of the ion-trajectories or variations in
ledge width) which may be of crucial importance in interpreting defect
images on the atomic level.

     The contrast from grain boundaries is more complex than that from
dislocations and is an area where computer simulation may find particularly
useful application.

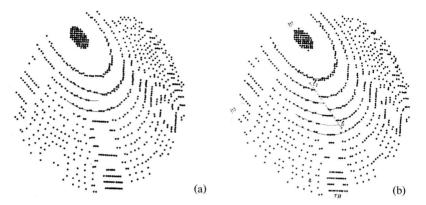

(a)                                                                              (b)

Fig. A1.4.  Examples of simulated contrast for perfect and dissociated dislocations in an
fcc metal. (Courtesy D. G. Brandon and A. J. Perry.)

## A1.4. The computer as an aid to indexing

Field-ion images are usually indexed with the aid of a standard projection and while these are readily available for cubic metals they are tedious
to construct for other materials. Sproul and Richman (1968) have produced
a program which generates projections for the hexagonal metals for any
selected $c/a$ ratio. Similarly, Morgan (1967) used a computer program to
predict the relative prominence and the angular relation of the poles in plates
of cementite.

## A1.5. The computer aided analysis of particle size and distribution

One dimension of a precipitate can be estimated from persistence during
a field evaporation sequence. However the persistence of particles of the
same size is a function of position with respect to the tip axis and the interpretation is conveniently done using a computer (Schwartz and Ralph,
1969). The method used is to record the position of a precipitate in a field-
ion image using an $X$-$Y$ trace reader, linked to a digital voltmeter and tape
punch. Data on the position and size of precipitates are analysed by the
computer taking details of the geometry into account. In this way the persistence of a large number of particles can be related to a mean particle diameter. A similar technique could probably be used to analyse a statistically
significant volume of data on point defects, for instance, and also assist in the
development of methods for utilising all the information on a micrograph.

# ELECTRON MICROSCOPE HOLDERS FOR VIEWING FIELD-ION TIPS

## A2.1. Introduction

In order to realise fully the potential of the field-ion microscope it is necessary to be able to correlate observations with those made by other techniques, and in particular by electron microscopy. Because of the limitations on specimen size imposed by the electron microscope, it is not always easy to examine specimens in such a way that they can be transferred backwards and forwards between electron and field-ion microscopes. Southon (1963) examined the field evaporated end forms of specimens by severing the end 2–3 mm of the tip and floating this on an electron microscope grid. This method is very simple if it is only desired to examine a few specimens, but the chances of damage to the specimen are very high, and it is not possible subsequently to re-examine the specimen in the field-ion microscope.

The adaptations which are necessary to the specimen holders of some common electron microscopes are described in this appendix. Microscopes with immersion objective lenses may be readily adapted to take wire specimens up to 1 cm long. For other electron microscopes with non-immersion objective lenses, it is only possible partly to overcome the difficulties outlined above without substantially modifying the actual microscope, and the imaging of specimens in the field-ion microscope *after* examination in the electron microscope remains difficult.

## A2.2. Design of specimen holders

### A2.2.1. MICROSCOPES WITH AN IMMERSION OBJECTIVE LENS

There are two high resolution electron microscopes widely distributed at the time of writing with an immersion objective lens. These are the Philips model EM300 (and EM200) and the GEC/AEI model EM6B (and EM801); the Philips has goniometer stage available and is thus the more useful of the two. The essential feature of an electron microscope with an immersion objective lens is that the pole piece of the lens is in two parts, one above and one below the specimen. The specimen itself is inserted by means of a simple rod. In the case of the EM6B and the high-resolution specimen stage of the EM300, the actual specimen holder is detached from the insertion rod during imaging to reduce the transmission of mechanical vibration through to the specimen.

Modified specimen holders for the high-resolution specimen stage of the EM300 are shown in figs. A2.1 and A2.2. In both cases specimens 1 cm or so long can be accommodated but axial rotation is not possible without removing the holder from the microscope, apart from the ±6° stereo tilt available in the microscope.

Modified specimen holders for the goniometer and rotation-tilt stages

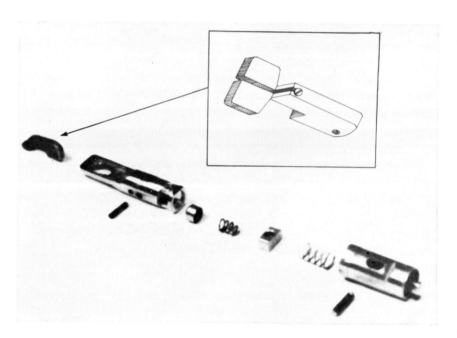

Fig. A2.1.   Modified specimen holder for the high resolution stage of a Philips EM300 (or EM200) electron microscope, enabling field-ion specimens up to 12 mm long to be examined: from left to right, specimen clamp; main body of holder; cap and springs.

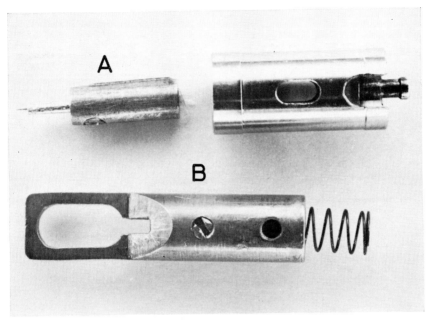

Fig. A2.2.   Alternative modification to the specimen holder for the high resolution stage
of the Philips EM300 (or EM200). The inner part (A) can be rotated relative to the outer
(B), but only by removing the holder from the microscope.
(Courtesy B. Loberg and H. Nordén.)

of the EM300 have been described by Loberg and Nordén (1969) and by
Smith and Bowkett (1968b). These have the tremendous advantage that it is
possible to rotate the specimen about its own long axis. Figure A2.3 shows
a modified holder for the goniometer stage (Loberg and Norden, 1969) based
on the standard tensile straining holder. Rotational motion is transferred by
means of a shaft supported at one end by a stainless steel ball bearing race
and an 0-ring seal. The other end of the shaft (B) ends with a screwdriver
blade, that engages in a groove in the part (C) actually holding the specimen
(F). The specimen is held in a chuck (D) and this means that it can be changed
without dismantling the holder. The spring (E) improves the stability of the
holder. Using such a holder the specimen can be rotated about its long axis
to an accuracy of $\pm 1°$; a small amount of tilt about an axis at right angles to this
is possible by using the height adjustment mechanism provided on the stage.

A2.2.2. MICROSCOPES WITH A NON-IMMERSION OBJECTIVE LENS.
A modification for an EM6G specimen holder is shown in fig. A2.4

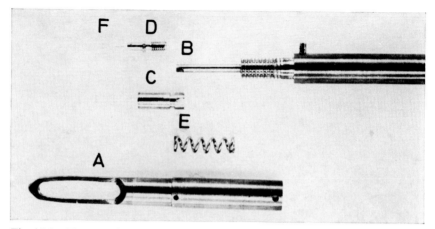

Fig. A2.3.   Front portion (A) of a rotation holder for the goniometer stage of the Philips EM300 (or EM200) electron microscope. The chuck (D) with the specimen (F) screws into (C) which can be rotated by the shaft (B). The spring (E) improves the stability of the holder. (Courtesy B. Loberg and H. Nordén.)

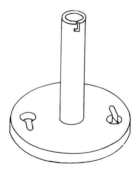

Fig. A2.4.   The simple modification necessary to the cap of the specimen holder for the EM6G electron microscope to enable field-ion specimens about 2.5 mm long to be examined.

which is largely self-explanatory. Two slits are milled at right angles to each other in a standard specimen holder cap. After the cap is in place on the specimen holder but before it is locked in position, a field-ion microscope specimen can be inserted through the slit, so that the tip comes in the centre of the specimen holder and the wire protrudes over the side. The cap is locked in place on two spring loaded studs, and the action of locking the cap on also locks the field-ion microscope specimen in place; the excess length of wire must then be cut off flush with the specimen holder. It will be appreciated that the maximum specimen length is then about 2.5 mm and it is still difficult or even impossible to re-examine the specimen in the field-ion microscope. However, this simple modification is a great improvement over the earlier method of floating specimens on grids. Similar modifications can be carried out to the standard specimen holders of other microscopes.

The maximum permissible specimen length with a non-immersion objective lens is limited by the diameter of the pole pieces. Loberg and Nordén (1969) have described a specimen holder suitable for use with the large (22 mm diameter) pole piece of the JEOL JEM7 (and JEM6) which will accept specimens with a maximum length of 10 mm.

An alternative method is to mount the specimen above the pole pieces of the objective lens rather than in the conventional position inside the pole

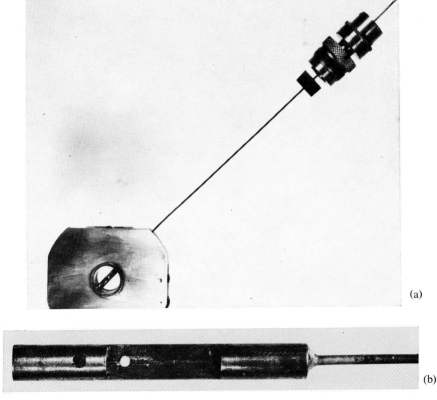

(a)

(b)

Fig. A2.5.    (a) Modified stage and specimen holder for the study of wire specimens in the Siemens Elmiskop I microscope. (b) Specimen holder enlarged.

piece. Figure A2.5 shows a holder and stage of this type for the Siemens Elmiskop 1 microscope. Some adjustments are necessary to the objective lens reference voltage in order to focus the specimen in its new plane. The specimen is held with vacuum grease, or spot welded in a spark machined groove, so that the end protrudes over the hole in the holder. The holder must be made of a non-magnetic material such as austenitic stainless steel and soldered onto a phosphor bronze wire which is sufficiently flexible to allow some translation of the specimen. Rotation of the specimen can be achieved with the external control knob shown in fig. A2.5. The disadvantage of this type of holder is that it has no airlock; specimen changing involves letting the microscope column down to air and takes about ten minutes.

# METHODS OF
# SPECIMEN PREPARATION

| Material | Electrolyte | Conditions | Additional remarks | Reference |
|---|---|---|---|---|
| Ir and its alloys e.g. Ir-W and Ir-Zr | aqueous $CrO_3$, 30% | 2–10 V ac | Rapid polish, useful for initial forming of the tip. Renew electrolyte at least once during polishing | Fortes (1968) |
| Ir | saturated aqueous $(NH_4)_2 CO_3$ | 2–30 V ac | Slow, but useful for final stages of tip preparation and back polishing | Fortes (1968) |
| Ir | aqueous KCN, 20% | 2–10 V ac | Slow | Ralph (1964) |
| Pt and its alloys | saturated aqueous $(NH_4)_2 CO_3$ | 2–30 V ac | Slow | Fortes (1968) |
| Pt and its alloys | aqueous KCN, 20% | 1–15 V ac | Slow | Ralph (1964) |
| Ru | aqueous KCl | 1.5–30 V ac | Concentration of solution not critical Ru counter electrode | Melmed and Klein (1966) |

*Continued Appendix 3*

| Material | Electrolyte | Conditions | Additional remarks | Reference |
|---|---|---|---|---|
| Ru | aqueous KOH | 1.5–30 V ac | As above | Melmed and Klein (1966) |
| Rh | aqueous KCN | 1 V ac | | Müller (1960) |
| Re | conc. $HNO_3$ | 10 V dc | | Müller (1960) |
| Re | equal volume mixture of 30% $H_2O_2$ and conc. $H_3PO_4$ | 3–9 V ac | | Melmed (1966) |
| Pd | 30% HCl 70% $HNO_3$ | 3 V ac | | Müller (1960) |
| Ni | 40% HCl | 1–2 V ac | | Müller (1960) |
| W | aqueous NaOH | 1–5 V ac | | Müller (1960) |
| W and its alloys e.g. W-Re | aqueous KCN, 20% | 1–10 V ac | Rapid polish | Ralph (1964) |
| Ta and its alloys and Nb and its alloys | 90 ml conc. $H_2SO_4$ and 10 ml HF, 40% | 1–10 V dc | Chill electrolyte, use stainless steel beaker or gold wire as cathode | Ralph (1964) Loberg and Nordén, private communication |
| Ta and its alloys and Nb and its alloys | 20 ml conc. $HNO_3$ 50 ml conc. $H_2SO_4$ 10 ml HF | 1–10 V dc | As above | Loberg and Nordén, private communication |
| Mo | aqueous KCN, 20% | 1–5 V ac | Rapid polish | Ralph (1964) |
| Mo | aqueous NaOH | 1–5 V ac | | Müller (1960) |
| Fe | 1% HCl | 0.5–1 V ac | Rapid polish | Müller (1960) Morgan (1967) |
| Fe and Fe-Mo-C | 133 ml glacial acetic acid and 25 g chromic oxide | 25 V dc | Slow, opaque, use "supervisor"; Morgan (1967) | Morgan et al. (1966) Gallot, private communication |

*Continued Appendix 3*

| Material | Electrolyte | Conditions | Additional remarks | Reference |
|---|---|---|---|---|
| Fe-Fe$_3$C | 9 ml ethyl alcohol 1 ml perchloric acid | 25 V dc | Slow | Morgan et al. (1966) |
| Zr | 10% HF | 10 V dc | | Müller (1960) |
| Ti | 40% HF | 4–12 V dc | | Müller (1960) |
| Be | conc. H$_3$PO$_4$ | 30–50 V dc | | Müller (1960) |
| Co | 10% HCl | 4–6 V dc 1–10 V ac | Rapid polish | Müller (1960) Turner (1967) |
| Au | equal volumes of conc. HNO$_3$ and conc. HCl | 1–10 V ac | | Müller (1960) |
| Cu | conc. H$_3$PO$_4$ | 1–5 V ac | | Müller (1960) |
| IrO$_2$ | | Oxidise field evaporated Ir tip in air or oxygen at $> 600°C$ | Polycrystalline specimens always obtained | Fortes and Ralph (1968b) |
| graphite | | Burn in coal gas flame | Not a regular end form | Müller (1960) |
| MgO | conc. H$_3$PO$_4$ and a trace of conc. HNO$_3$ | Chemical polish at $125°C$ using two layer technique | | Taylor and Turner (1968), private communication |
| SiC | | Burn in oxy-propane flame; remove residual SiO$_2$ in HF | | Smith (1969a) |
| Si and Ge | 15 ml acetic acid 15 ml HF, 40% 25 ml conc. HNO$_3$ | Chemical polish | Quite rapid | Müller (1960) Arthur (1964) |

*Continued Appendix 3*

| Material | Electrolyte | Conditions | Additional remarks | Reference |
|---|---|---|---|---|
| TiC | 15 ml acetic acid | 3–15 V ac or dc | Ta counter electrode | Smith (1968) Meakin (1968) |
| TiC, NbC and TaC | 15 ml HF, 40% 25 ml conc. $HNO_3$, trace of HBr | | | |
| ZrC | conc. $HNO_3$ | 3–10 V dc | Stainless steel cathode | Smith (1968) |
| $UO_2$ | 310 ml conc. $H_3PO_4$, 38 g $CrO_3$ 67 ml conc. $H_2SO_4$ 120 ml water | 60 V dc | Slow | Morgan (1968) |
| TiN | 40% HF | 1–6 V dc | | Müller (1960) |
| WC and $W_2C$ | | carburise field evaporated tungsten tip *in situ* | | French and Richman (1968) |
| WC | 10% aqueous NaOH | | | Meakin (1968) |

# TABLES OF INTERPLANAR AND INTERATOM DISTANCES

**A4.1. Table of interatom distances (all in Å) for some hexagonal metals**

|            | $\langle 11\bar{2}0\rangle \equiv$ $\langle 100\rangle$ or $\langle 110\rangle$ | $\langle 2\bar{1}\bar{1}3\rangle \equiv$ $\langle 101\rangle$ or $\langle 111\rangle$ | [0001] $\equiv$[001] | $\langle 10\bar{1}2\rangle$ $\equiv\langle 212\rangle$ | $\langle 10\bar{1}1\rangle$ $\equiv\langle 211\rangle$ | $\langle 11\bar{2}6\rangle$ $\equiv\langle 112\rangle$ | $\langle 01\bar{1}0\rangle$ $\equiv\langle 120\rangle$ |
|------------|------|------|------|------|------|------|------|
| Cobalt     | 2.51 | 4.78 | 4.07 | 9.23 | 5.95 | 8.52 | 4.34 |
| Hafnium    | 3.21 | 6.63 | 5.09 | 11.6 | 7.53 | 10.7 | 5.54 |
| Osmium     | 2.73 | 5.11 | 4.32 | 9.85 | 6.41 | 9.06 | 4.73 |
| Rhenium    | 2.76 | 5.24 | 4.46 | 10.1 | 6.54 | 9.33 | 4.78 |
| Ruthenium  | 2.70 | 5.06 | 4.28 | 9.76 | 6.19 | 8.98 | 4.68 |
| Titanium   | 2.95 | 5.53 | 4.68 | 10.7 | 6.93 | 9.82 | 5.10 |
| Yttrium    | 3.67 | 6.89 | 5.82 | 13.3 | 8.62 | 12.2 | 6.36 |
| Zirconium  | 3.23 | 6.60 | 5.13 | 11.7 | 7.59 | 10.8 | 5.59 |

## A4.2. Interplanar spacings (all in Å) for some metals with the hexagonal structure

| | $c/a$ ratio | $\{0001\}$ | $\{11\bar{2}0\}$ | $\{10\bar{1}0\}$ | $\{10\bar{1}1\}$ | $\{11\bar{2}1\}$ | $\{12\bar{3}0\}$ | $\{10\bar{1}2\}$ | $\{11\bar{2}2\}$ | $\{20\bar{2}1\}$ |
|---|---|---|---|---|---|---|---|---|---|---|
| Cobalt | 1.624 | 2.03 | 1.26 | 1.45, 0.72 | 1.60, 0.32 | 0.60 | 0.55, 0.27 | 0.99, 0.49 | 1.07 | 0.88, 0.18 |
| Hafnium | 1.587 | 2.54 | 1.61 | 1.85, 0.93 | 2.03, 0.41 | 0.76 | 0.70, 0.35 | 1.25, 0.62 | 1.35 | 1.11, 0.22 |
| Osmium | 1.580 | 2.16 | 1.37 | 1.58, 0.79 | 1.73, 0.35 | 0.65 | 0.60, 0.30 | 1.06, 0.53 | 1.15 | 0.95, 0.19 |
| Rhenium | 1.615 | 2.23 | 1.38 | 1.59, 0.80 | 1.75, 0.35 | 0.66 | 0.60, 0.30 | 1.09, 0.54 | 1.17 | 0.97, 0.19 |
| Ruthenium | 1.584 | 2.14 | 1.35 | 1.56, 0.78 | 1.72, 0.34 | 0.65 | 0.60, 0.30 | 1.05, 0.53 | 1.14 | 0.95, 0.19 |
| Titanium | 1.587 | 2.34 | 1.48 | 1.70, 0.85 | 1.87, 0.37 | 0.70 | 0.64, 0.32 | 1.15, 0.57 | 1.25 | 1.03, 0.21 |
| Yttrium | 1.587 | 2.91 | 1.84 | 2.11, 1.05 | 2.32, 0.46 | 0.87 | 0.79, 0.40 | 1.43, 0.72 | 1.54 | 1.27, 0.25 |
| Zirconium | 1.589 | 2.57 | 1.62 | 1.87, 0.93 | 2.05, 0.41 | 0.77 | 0.70, 0.35 | 1.26, 0.63 | 1.37 | 1.11, 0.22 |

*Note*: for planes $(hkil)$ where $2h+4k+3l$ is not an integral multiple of 6, except some planes where it takes the value $6N-3$ ($N$ is an integer), the interplanar spacing alternates between two values. The spacing of planes not listed above can be found using the method given by Nicholas (1965) p. 21.

A4.3. Table of interatom distances (all in Å) for some cubic metals

| | Lattice type | ⟨100⟩ | ⟨110⟩ | ⟨111⟩ | ⟨120⟩ | ⟨112⟩ | ⟨122⟩ | ⟨130⟩ | ⟨113⟩ | ⟨230⟩ | ⟨231⟩ |
|---|---|---|---|---|---|---|---|---|---|---|---|
| Cobalt | Cubic F* | 3.55 | 2.51 | 6.14 | 7.92 | 4.34 | 10.6 | 5.60 | 11.8 | 12.8 | 6.63 |
| Gold | Cubic F | 4.08 | 2.88 | 7.06 | 9.12 | 4.98 | 12.2 | 6.45 | 13.5 | 14.7 | 7.63 |
| Iridium | Cubic F | 3.84 | 2.71 | 6.65 | 8.58 | 4.69 | 11.5 | 6.07 | 12.7 | 13.8 | 7.19 |
| Iron | Cubic I | 2.87 | 4.05 | 2.48 | 6.41 | 7.01 | 8.58 | 9.05 | 4.75 | 10.3 | 10.7 |
| Molybdenum | Cubic I | 3.15 | 4.45 | 2.72 | 7.03 | 7.70 | 9.43 | 9.94 | 5.22 | 11.3 | 11.8 |
| Nickel | Cubic F | 3.52 | 2.49 | 6.10 | 7.88 | 4.31 | 10.6 | 5.57 | 11.7 | 12.7 | 6.59 |
| Niobium | Cubic I | 3.30 | 4.66 | 2.85 | 7.38 | 8.09 | 9.90 | 10.4 | 5.46 | 11.9 | 12.3 |
| Palladium | Cubic F | 3.89 | 2.75 | 6.73 | 8.70 | 4.75 | 11.6 | 6.15 | 12.9 | 14.0 | 7.28 |
| Platinum | Cubic F | 3.92 | 2.77 | 6.80 | 8.78 | 4.80 | 11.8 | 6.20 | 13.0 | 14.1 | 7.33 |
| Rhodium | Cubic F | 3.80 | 2.69 | 6.59 | 8.50 | 4.65 | 11.4 | 6.00 | 12.6 | 13.7 | 7.11 |
| Tantalum | Cubic I | 3.30 | 4.67 | 2.86 | 7.38 | 8.09 | 9.90 | 10.4 | 5.48 | 11.9 | 12.4 |
| Tungsten | Cubic I | 3.16 | 4.47 | 2.74 | 7.08 | 7.75 | 9.49 | 10.0 | 5.24 | 11.4 | 11.8 |
| Vanadium | Cubic I | 3.04 | 4.30 | 2.62 | 6.79 | 7.43 | 9.12 | 9.61 | 5.04 | 11.0 | 11.4 |

* High temperature allotrop which transforms to hexagonal structure below about 440 °C (may still exist, however, at considerably lower temperatures).

## A4.4. Table of interplanar spacings (all in Å) for some body centred cubic metals

| | Iron | Molyb-denum | Niobium | Tantalum | Tungsten | Vanadium |
|---|---|---|---|---|---|---|
| Lattice parameter | 2.866 | 3.148 | 3.302 | 3.304 | 3.165 | 3.038 |
| **Plane** | | | | | | |
| 110 | 2.027 | 2.225 | 2.334 | 2.335 | 2.239 | 2.149 |
| 200 | 1.433 | 1.574 | 1.651 | 1.652 | 1.583 | 1.519 |
| 112 | 1.170 | 1.285 | 1.348 | 1.349 | 1.292 | 1.241 |
| 130 | 0.906 | 0.995 | 1.043 | 1.044 | 1.001 | 0.961 |
| 222 | 0.828 | 0.909 | 0.953 | 0.954 | 0.914 | 0.878 |
| 231 | 0.766 | 0.841 | 0.882 | 0.883 | 0.846 | 0.812 |
| 240 | 0.641 | 0.704 | 0.738 | 0.739 | 0.708 | 0.679 |
| 244 | 0.478 | 0.525 | 0.550 | 0.551 | 0.528 | 0.507 |
| 226 | 0.432 | 0.475 | 0.498 | 0.498 | 0.477 | 0.458 |
| 460 | 0.398 | 0.437 | 0.458 | 0.458 | 0.439 | 0.422 |

## A4.5. Table of interplanar spacings (all in Å) for some face centred cubic metals

| | Cobalt* | Gold | Iridium | Nickel | Palladium | Platinum | Rhodium |
|---|---|---|---|---|---|---|---|
| Lattice parameter | 3.550 | 4.078 | 3.838 | 3.524 | 3.890 | 3.924 | 3.804 |
| **Plane** | | | | | | | |
| 111 | 2.050 | 2.355 | 2.217 | 2.034 | 2.246 | 2.266 | 2.196 |
| 200 | 1.776 | 2.039 | 1.919 | 1.762 | 1.945 | 1.962 | 1.902 |
| 220 | 1.256 | 1.442 | 1.357 | 1.246 | 1.376 | 1.387 | 1.345 |
| 113 | 1.070 | 1.230 | 1.158 | 1.064 | 1.174 | 1.184 | 1.148 |
| 240 | 0.794 | 0.912 | 0.859 | 0.790 | 0.870 | 0.877 | 0.850 |
| 224 | 0.725 | 0.833 | 0.784 | 0.720 | 0.794 | 0.802 | 0.776 |
| 244 | 0.592 | 0.679 | 0.640 | 0.588 | 0.649 | 0.654 | 0.634 |
| 260 | 0.562 | 0.645 | 0.607 | 0.557 | 0.615 | 0.621 | 0.601 |
| 460 | 0.493 | 0.566 | 0.532 | 0.489 | 0.539 | 0.544 | 0.527 |
| 462 | 0.475 | 0.545 | 0.513 | 0.471 | 0.520 | 0.524 | 0.508 |

* High temperature allotrop which transforms to hexagonal structure below about 440 °C (may still exist, however, at considerably lower temperatures).

APPENDIX **5**

# THE INDEXING
# OF MICROGRAPHS

This appendix contains a series of comparative micrographs and stereo-graphic projections. Only low index poles are marked on the stereograms; the position of higher index poles can be found by interpolation using the zone-adding rule.

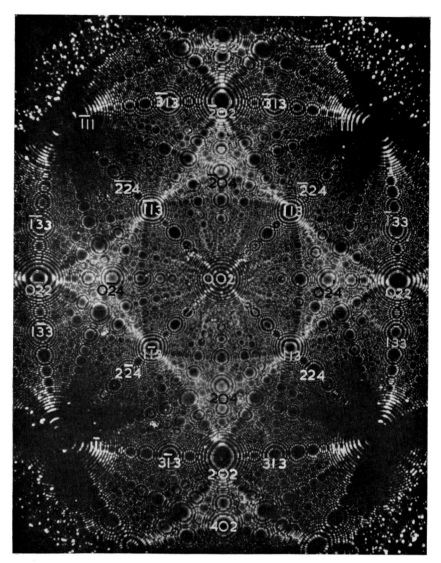

Fig. A5.1a.   Platinum: micrograph of (001) oriented specimen. (Courtesy E. W. Müller.)

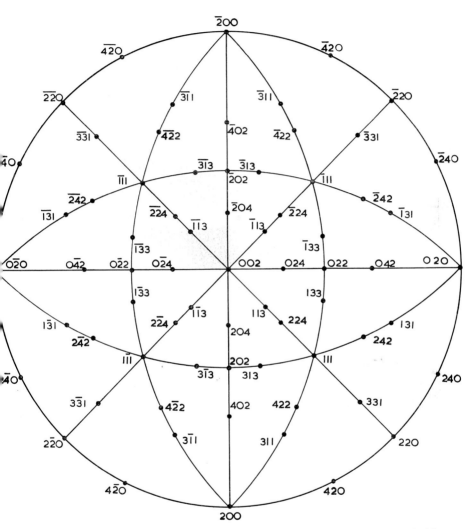

Fig. A5.1b.  Stereographic projection of the cubic lattice on the (001) plane, marked in accordance with the relative prominence rule for fcc.

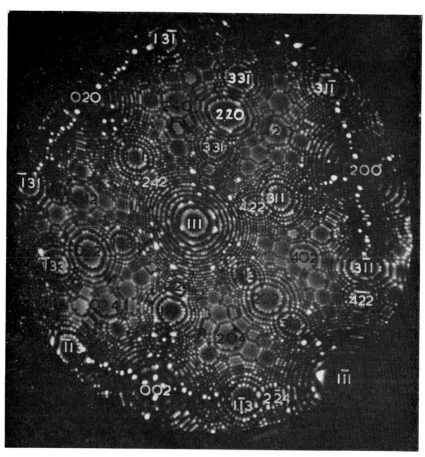

Fig. A5.2a.   Iridium: micrograph of (111) oriented specimen. (Courtesy T. F. Page.)

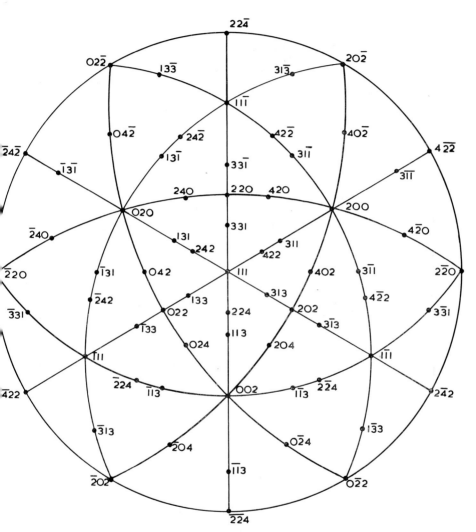

Fig. A5.2b.   Stereographic projection of the cubic lattice on the (111) plane, marked in accordance with the relative prominence rule for fcc.

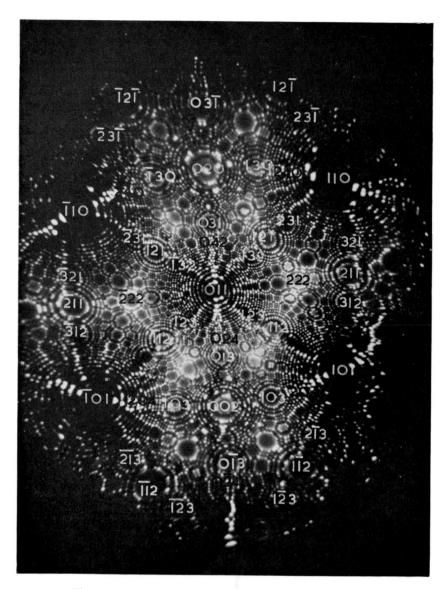

Fig. A5.3a.   Tungsten: micrograph of (011) oriented specimen.

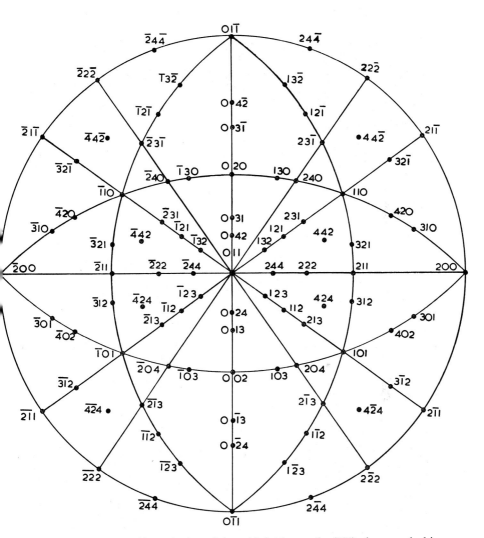

Fig. A5.3b.  Stereographic projection of the cubic lattice on the (011) plane, marked in
accordance with the relative prominence rule for bcc.

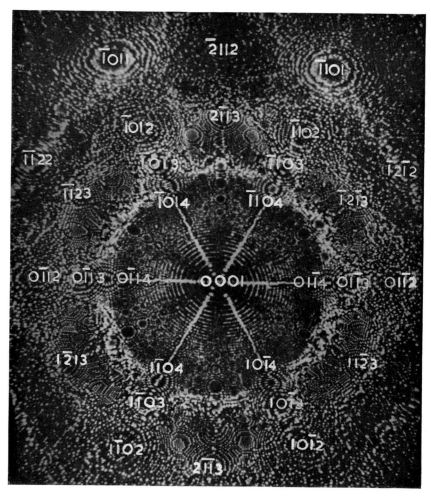

Fig. A5.4a.    Rhenium ($c/a = 1.62$): micrograph of (0001) oriented specimen.
(Courtesy E. W. Müller.)

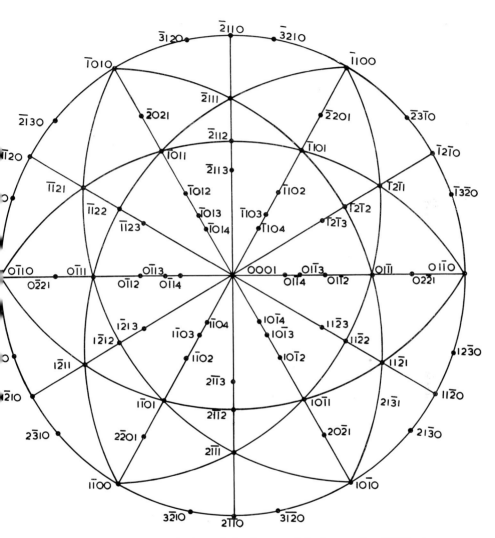

Fig. A5.4b.  Stereographic projection of the hexagonal lattice on the (0001) plane ($c/a = 1.633$; the ideal ratio for the cph structure).

# EQUIPMENT SUPPLIERS

| Suppliers | Makers of |
|---|---|
| Air Products and Chemicals Inc., Advanced Products Department, Allentown, Pennsylvania, U.S.A. | Miniature gas liquifiers ("Cryotips") |
| Bendix Corporation, 605 3rd Avenue, New York, N.Y., U.S.A. | Channel plates and single particle detectors |
| Brandenburg Ltd., 139, Sanderstead Road, South Croydon, Surrey, England | Power supplies |
| Edwards High Vacuum International Ltd., Manor Royal, Crawley, Sussex, England | Vacuum equipment |
| E.M.I. Electronics Ltd., Hayes, Middlesex, England | Image intensifiers |
| G.E.C.–A.E.I. Ltd., P.O. Box 1, Harlow, Essex, England | Electron microscopes and image intensifiers |
| Hymatic Engineering Co. Ltd., 40 Broadway, London, S.W.1, England | Miniature gas liquifiers |
| Jackson and Church Electronics Company, 1127 South Patric Drive, Satellite Beach, Florida, U.S.A. | Field-ion microscopes |
| Levy West Ltd., Bush Fair, Harlow, Essex, England | Phosphors |
| M.R.C., Route 303, Orangeburg, New York, U.S.A. | Specimen wires |
| Mosaic Corporation, New York, U.S.A. | Fibre optic windows |
| Mullard Ltd., New Road, Mitcham, Surrey, England | Channel plates |
| Polaron Instruments Ltd., Deviljem House, Finchley, London N.3, England | Stereographic projections |
| Philips Co., Eindhoven, Holland | Electron microscopes |
| R.C.A., Lancaster, Pa., U.S.A. | Image intensifiers |
| Shackman, Mineral Lane, Chesham, Bucks., England | Cameras |
| Twentieth Century Electronics Ltd., Centronics Works, King Henry's Drive, New Addington, Croydon, Surrey, England | Field-ion microscopes |
| Vacuum Generators Ltd., Charlwood Road, East Grinstead, Sussex, England | Field-ion microscopes |
| Velonex Co., Santa Clara, California, U.S.A. | Pulse generator |
| Wray Ltd., Ashgrove Road, Bromley, Kent, England | Lenses |

# REFERENCES

Amelinckx, S. and W. Dekeyser, 1959, Solid State Phys. **8**, 325.
Anderson, J. S., 1967, Record of the Australian Academy of Science **1**, 109.
Arthur, J. H., 1964, J. Phys. Chem. Solids **25**, 583.
Attardo, M. J. and J. M. Galligan, 1966, Phys. Stat. Sol. **16**, 449.
Attardo, M. J., J. M. Galligan and J. Sadovsky, 1966, J. Sci. Inst. **43**, 607.
Attardo, M. J. and J. M. Galligan, 1967, 14th Field Emission Symposium, Washington.
Attardo, M. J., J. M. Galligan and J. G. Y. Chow, 1967, Phys. Rev. Letters **19**, 73.
Barofsky, D. F. and E. W. Müller, 1968, Surface Science **10**, 177.
Bassett, D. W., 1965, Proc. Roy. Soc. **A268**, 191.
Bassett, D. W., 1966, Disc. Faraday Soc. **41**, 65.
Bassett, D. W. and M. J. Parsley, 1969, Nature **221**, 1046.
Becker, J. A., 1951, Bell Systems Tech. J. **30**, 907.
Beckey, H. D., H. Krone and F. W. Rollgen, 1968, J. Sci. Instr., Series 2, **1**, 118.
Bell, A. E., E. C. Cooper and R. J. Adams, 1966, 13th Field Emission Symposium, Cornell University.
Boudreaux, D. S. and P. H. Cutler, 1966a, Surface Science **5**, 230.
Boudreaux, D. S. and P. H. Cutler, 1966b, Phys. Rev. **149**, 170.
Bowden, F. P. and L. T. Chadderton, 1962, Proc. Roy, Soc. **A269**, 143.
Bowkett, K. M., 1966, Ph. D. thesis, University of Cambridge.
Bowkett, K. M., J. J. Hren and B. Ralph, 1964, Third European Regional Conference on Electron Microscopy, Prague, 191.
Bowkett, K. M., L. T. Chadderton, H. Nordén and B. Ralph, 1965, Phil. Mag. **11**, 651.
Bowkett, K. M. and B. Ralph, 1966, J. Sci. Instr. **43**, 703.
Bowkett, K. M., L. T. Chadderton, H. Nordén and B. Ralph, 1967, Phil. Mag. **15**, 415.
Bowkett, K. M. and B. Ralph, 1969, Proc. Roy. Soc. **A 312**, 51.
Bowkett, K. M., J. Gallott, D. A. Smith, G. D. W. Smith and W. A. Soffa, 1969, Intern. Conf. on Quantitative Metallography, Haifa. Eds. Brandon and Rosen (Israel University Press, Tel Aviv, 1970) p. 511.
Brandon, D. G., 1963, Brit. J. Appl. Phys. **14**, 474.
Brandon, D. G., 1964a, Surface Science **2**, 495.
Brandon, D. G., 1964b, J. Sci. Instr. **41**, 373.
Brandon, D. G., 1964c, 11th Field Emission Symposium, Cambridge, England.
Brandon, D. G., 1965, Surface Science **3**, 1.

Brandon, D. G., 1966a, Field-Ion Microscopy, Eds. Hren and Ranganathan (Plenum Publishing Corporation, New York, 1968).

Brandon, D. G., 1966b, Phil. Mag. **14**, 803.

Brandon, D. G., 1966c, Surface Science **5**, 137.

Brandon, D. G., 1966d, Phil. Mag. **13**, 1085.

Brandon, D. G., 1966e, Acta Met. **14**, 1479.

Brandon, D. G., 1966f, J. Sci. Instr. **43**, 708.

Brandon D. G., M. J. Southon and M. S. Wald, 1961, Proc. Int. Conf., Berkeley Castle, Gloucestershire, England (Butterworth, London) p. 113.

Brandon, D. G. and M. S. Wald, 1961, Phil. Mag. **6**, 1035.

Brandon, D. G., B. Ralph, S. Ranganathan and M. S. Wald, 1964, Acta Met. **12**, 813.

Brandon, D. G., S. Ranganathan and D. S. Whitmell, 1964, Brit. J. Appl. Phys. **15**, 55.

Brandon, D. G. and A. J. Perry, 1967a, Phil. Mag. **16**, 119.

Brandon, D. G. and A. J. Perry, 1967b, Phil. Mag. **16**, 131.

Brenner, S. S., 1962, A.S.M. Publication on Metal Surfaces.

Brenner, S. S., 1965, High Temperature, High Resolution Metallography, Eds. Aaronson and Ansell (Gordon and Breach, New York).

Brenner, S. S., 1966, 13th Field Emission Symposium, Cornell University.

Brenner, S. S. and J. T. McKinney, 1968, Appl. Phys. Letters **13**, 29.

Brooks, H., 1955, Impurities and Imperfections, A.S.M.

Buswell, J. T., 1970, Ph.D. thesis, University of London; Phil. Mag. **21**, 357.

Caspary, E. K. and E. Krautz, 1965, Z. Naturforschg. **19a**, 591.

Chang, R., 1967, Phil. Mag. **16**, 1021.

Cinquana, L. and J. D. Meakin, 1967, Tech. Rept., Franklin Inst., P - C1780 - 1.

Cohen, J. B., R. Hinton, K. Lay and S. Sass, 1962, Acta Met. **10**, 894.

Cottrell, A. H., 1964, Theory of Crystal Dislocations (Blackie, London).

Cottrell, A. H., 1965, Mechanical Properties of Matter (Wiley, New York).

Cranstoun, G. K. L., 1968a, 4th Symposium on Photo-electric Image Devices, Imperial College, London.

Cranstoun, G. K. L., 1968b, 15th Field Emission Symposium, Bonn.

Cranstoun, G. K. L. and Anderson. J. S., 1966, 13th Field Emission Symposium, Cornell University.

Crussard, C., 1961, C. R. Acad. Sci. (Paris) **9**, 273.

Drechsler, M. and P. Wolf, 1958, Proc. 4th Int. Conf. on Electron Microscopy, Berlin (Springer, Berlin) **1**, 835.

Dubroff, W., 1967, Ph. D. thesis, University of Columbia, New York.

Dubroff, W. and E. S. Machlin, 1968, Acta Met. **16**, 1313.

Duke, C. B. and M. E. Alferieff, 1967, J. Chem. Phys. **46**, 938.

Dyke, W. P. and W. W. Dolan, 1956, Adv. Electron. and Electron Phys. **8**, 89.

Dyke, W. P., J. K. Trolan, W. W. Dolan and G. Barnes, 1953, J. Appl. Phys. **24**, 570.

Ehrlich, G. and F. G. Hudda, 1960, J. Chem. Phys. **33**, 1253.

Ehrlich, G. and F. G. Hudda, 1962, J. Chem. Phys. **36**, 3233.

Ehrlich, G. and F. G. Hudda, 1966, J. Chem. Phys., **44**, 1039.

Elvin, C. D., 1967, Phil. Mag. **16**, 35.

Ernst, L. G., 1966, Phys. Stat. Sol. **14**, K107.

Ernst, L. G., 1967, Phys. Stat. Sol. **19**, 89.

Eshelby, J. D. and A. N. Stroh, 1951, Phil. Mag. **42**, 1401.

Eyring, C. F., S. S. Mackeown and R. A. Millikan, 1928, Phys. Rev. **31**, 900.

Fasth, J. E., B. Loberg and H. Nordén, 1967, J. Sci. Instr. **44**, 1044.

Faulkner, R. G., P. J. Turner and B. Ralph, 1968, 4th European Regional Conf. on Electron Microscopy, Rome, p. 445.

Feldman, U. and R. Gomer, 1966, J. Appl. Phys. **37**, 2380.

Forbes, R. G. and M. J. Southon, 1966, 13th Field Emission Symposium, Cornell University.

Fortes, M. A., 1968, Ph. D. thesis, Cambridge University.

Fortes, M. A. and B. Ralph, 1967, Acta Met. **15**, 707.

Fortes, M. A. and B. Ralph, 1968a, Phil. Mag. **18**, 787.

Fortes, M. A. and B. Ralph, 1968b, Proc. Roy. Soc. A**307**, 431.

Fortes, M. A. and B. Ralph, 1969a, Phil. Mag. **19**, 181.

Fortes, M. A. and B. Ralph, 1969b, Cambridge University, internal report.

Fortes, M. A. and D. A. Smith, 1970, J. Appl. Phys., to be published.

Foxall, R. A., M. S. Duesbery and P. B. Hirsch, 1967, Can. J. Phys. **45**, 607.

Frank, F. C., 1950, Symp. on Plastic Deformation of Crystalline Solids, Carnegie-Mellon, p. 69.

Frank, F. C., 1951, Phil. Mag. **42**, 809.

French, R. D. and M. H. Richman, 1968, Phil. Mag. **18**, 471.

Galligan, J. M. and M. J. Attardo, 1966, 13th Field Emission Symposium, Cornell University.

Gillott, L., 1968, Ph. D. thesis, University of Cambridge.

Gillott, L. and M. J. Southon, 1967, Institute of Physics meeting on Field-ion Microscopy, Cambridge.

Gilman, J. J., 1962, J. Appl. Phys. **33**, 2703.

Gold, E. and E. S. Machlin, 1968, Phil. Mag. **18**, 453.

Gomer, R., 1961, Field Emission and Field Ionisation (Harvard University Press, Cambridge, Mass.).

Gomer, R. and L. W. Swanson, 1963, J. Chem. Phys. **38**, 1613.

Good, R. H. and E. W. Muller, 1956, Handbuch der Physik **21**, 176.

Goux, C., 1961, Mem. Sci. Rev. Met. **58**, 661.

Goux, C., 1963, Acta Met. **11**, 111.

Hirsch, P. B., 1968, Proc. Int. Conf. on Strength of Metals and Alloys, Supp. Trans. Jap. Inst. Met. **9**, xxx.

Hirsch, P. B., A. Howie, R. B. Nicholson, D. W. Pashley and M. J. Whelan, 1965, Electron Microscopy of Thin Crystals (Butterworths, London).

Holland, B. W., 1963, Phil. Mag. **8**, 87.

Honeycombe, R. W. K., 1968, The Plastic Deformation of Metals (Arnold, London).

Hren, J. J., 1965, Acta Met. **13**, 479.

Hudson, J. A., 1969, Ph. D. thesis, Cambridge University.

Hudson, J. A., R. S. Nelson and B. Ralph, 1968, Phil. Mag. **18**, 839; AERE, Harwell, Report R5747.

Inghram, M. and R. Gomer, 1954, J. Chem. Phys. **22**, 1279.

Jason, A. J., 1967, Phys. Rev. **15**, 266.

Jason, A. J., R. P. Burns, A. C., Parr and M. Inghram, 1966, J. Chem. Phys. **44**, 4351.

Jeanotte, D. and J. M. Galligan, 1970, Acta Met. **18**, 71.

Jones, J. P., 1966, Nature **211**, 479.

Kelly, A., 1966, Strong Solids (Oxford University Press).

Kelly, A. and R. B. Nicholson, 1963, Prog. Mat. Sci. **10**, 151.

Klipping, G. and R. Vanselow, 1967, Z. Phys. Chem. **52**, 196.

Knor, Z. and E. W. Müller, 1968, Surface Science **10**, 21.

Kraftmakher, A. and P. G. Strelkov, 1963, Soviet Phys. Solid. State **4**, 1662.

Kroupa, F., 1963, Phys. Stat. Sol. **3**, K391.

Kroupa, F. and V. Vitek, 1964, Czech. J. Phys. B**14**, 337.

Langdon, T. G. and J. E. Dorn, 1968, Phil. Mag. **17**, 999.

LeFevre, B. G., H. Grenga and B. Ralph, 1968, Phil. Mag. **18**, 1127.

Lewis, R. and R. Gomer, 1969, Appl. Phys. Letters **15**, 384.

Loberg, B. and H. Nordén, 1967, Phil. Mag. **16**, 1147.
Loberg, B. and H. Nordén, 1969, Arkiv for Fysik **39**, 383.
Machlin, E. S., 1967, Trans. A.S.M. **60**, 260.
Marcinkowski, M. J., 1964, Electron Microscopy and Strength of Materials, Eds. Thomas and Washburn (Interscience, New York).
McLane, S. B., E. W. Müller and O. Nishikawa, 1964, Rev. Sci. Instr. **35**, 1297.
McNeill, J. F., 1968, 15th Field Emission Symposium, Bonn.
Meakin, J. D., 1968, Phil. Mag. **17**, 865.
Meakin, J. D., A. Lawley and R. C. Koo, 1964, Appl. Phys. Letters **5**, 133.
Melmed, A. J., 1966a, Surface Science **5**, 359.
Melmed, A. J., 1966b, Field-ion microscopy, Eds. Hren and Ranganathan (Plenum Publishing Corporation, New York, 1968).
Melmed, A. J., 1967, Surface Science **8**, 191.
Melmed, A. J. and R. Klein, 1966, J. Less Common Metals **10**, 225.
Meyrick, G., 1967, J. Less Common Metals **12**, 242.
Montagu-Pollock, H. M. and T. N. Rhodin, 1966, J. Sci. Inst. **43**, 667.
Montagu-Pollock, H. M., T. N. Rhodin and M. J. Southon, 1968, Surface Science **12**, 1.
Moore, A. J. W., 1962, J. Phys. Chem. Solids **23**, 907.
Moore, A. J. W., 1967, Phil Mag. **16**, 723.
Moore A. J. W. and D. G. Brandon, 1968, Phil. Mag. **18**, 679.
Moore, A. J. W. and S. Ranganathan, 1967, Phil. Mag. **16**, 739.
Moore, A. J. W. and J. A. Spink, 1968, Surface Science **12**, 479.
Morgan, R., 1967a, Ph. D. thesis, Cambridge University.
Morgan, R., 1967b, J. Sci. Instr. **44**, 808.
Morgan, R., R. G. Faulkner and B. Ralph, 1966, J.I.S.I. **204**, 943.
Morgan, R. and B. Ralph, 1968, J.I.S.I. **206**, 1138.
Mott, N. F., 1948, Proc. Phys. Soc. **60**, 391.
Müller, E. W., 1949, Z. für Physik **126**, 462.
Müller, E. W., 1955, Ann. Meeting of El. Mic. Soc. of America, Pennsylvania State University.
Müller, E. W., 1956, Z. Naturforsch. **11**a, 88; J. Appl. Phys. **27**, 474.
Müller, E. W., 1957, J. Appl. Phys. **28**, 6.
Müller, E. W., 1958a, Proc. 4th Int. Conf. Electron Microscopy, Berlin (Springer, Berlin) **1**, 820.
Müller, E. W., 1958b, Acta Met. **6**, 620.
Müller, E. W., 1959, Reactivity of Solids, Proc. Intern. Symposium Reactivity of Solids, Ed. de Boer et al. (Elsevier, Amsterdam, 1960) p. 691.
Müller, E. W., 1960, Adv. Electronics and Electron Physics **13**, 83.
Müller, E. W., 1962, Proc. Int. Conf. on Crystal Lattice Defects, Kyoto, J. Phys. Soc. Japan **18**, Supp. II, 1963, p. 1.
Müller, E. W., 1964, Surface Science **2**, 484.
Müller, E. W., 1965, Science **149**, 591.
Müller, E. W., 1967a, Z. für Phys. Chem. **53**, 1.
Müller, E. W., 1967b, Annual Review of Physical Chemistry **18**, 35.
Müller, E. W. and E. Bahadur, 1956, Phys. Rev. **102**, 624.
Müller, E. W., S. Nakamura, O. Nishikawa and S. B. McLane, 1965, J. Appl. Phys. **36**, 2496.
Müller, E. W. and K. D. Rendulic, 1966, 13th Field Emission Symposium, Cornell University.
Müller, E. W., J. A. Panitz and S. B. McLane, 1968, Rev. Sci. Instr. **39**, 83.
Mulson, J. F. and E. W. Müller, 1959, 6th Field Emission Symposium, Washington.
Mulson, J. F. and E. W. Müller, 1963, J. Chem. Phys. **38**, 2615.

Murr, L. E. and M. C. Inman, 1965, Phys. Stat. Sol. **10**, 441.

Nadakavukaren, J. J. and F. Hutchinson, 1966, 13th Field Emission Symposium, Cornell University.

Nakamura, S. and E. W. Müller, 1965, J. Appl. Phys. **36**, 3634.

Nelson, R. S., 1963, Phil. Mag. **8**, 693.

Newman, R. W., 1967, Ph. D. thesis, University of Florida, Gainesville; 1968, Scripta Met. **2**, 69.

Newman, R. W., B. G. LeFevre and J. J. Hren, 1966, 13th Field Emission Symposium, Cornell University.

Newman, R. W. and J. J. Hren, 1967, Surface Science **8**, 373.

Newman, R. W., R. Sanwald and J. J. Hren, 1967, J. Sci. Instr. **44**, 828.

Nicholas, J. F., 1965, An Atlas of Models of Crystal Surfaces (Gordon and Breach, New York).

Nordén, H., 1969, Thesis, Chalmers Technical University, Gothenburg.

Page, T. F., 1970, Ph. D. thesis, Cambridge University.

Parsley, M. J., 1969, Ph. D. thesis, University of London.

Partridge, P. G., 1968, Metallurgical Reviews **118**, 169.

Pashley, D. W., 1965, Rept. Prog. Phys. **28**, 291.

Perry, A. J. and D. G. Brandon, 1969, Surface Science **14**, 423.

Petroff, P., 1967, Ph. D. thesis, University of California, Lawrence Radiation Lab., Berkeley, UCRL-17633.

Petroff, P. and J. Washburn, 1967, University of California, Lawrence Radiation Lab., Berkeley, UCRL-17347.

Pimbley, W. T. and R. M. Ball, 1965, Rev. Sci. Inst. **36**, 225.

Pimbley, W. T., C. A. Speicher, M. J. Attardo, J. M. Galligan and S. S. Brenner, 1966, 13th Field Emission Symposium, Cornell University.

Plummer, E. W., 1967, Ph. D. thesis, Cornell University.

Plummer, E. W. and T. N. Rhodin, 1968, J. Chem. Phys. **49**, 3479.

Raghaven, N.V., 1967, Ph. D. thesis, University of California.

Ralph, B., 1964, Ph. D. thesis, Cambridge University.

Ralph, B., 1966, Field-Ion Microscopy, Eds. Hren and Ranganathan (Plenum Publishing Corporation, New York, 1968).

Ralph, B. and D. G. Brandon, 1963, Phil. Mag. **8**, 919.

Ralph, B. and D. G. Brandon, 1964, Journées Internationales des Applications du Cobalt, Bruxelles, p. 1.

Ranganathan, S., 1964, 3rd European Regional Conference on Electron Microscopy, Prague (Publishing House of the Czech. Academy of Science, Prague).

Ranganathan, S., 1965, Ph. D. thesis, Cambridge University.

Ranganathan, S., 1966a, J. Appl. Phys. **37**, 4346.

Ranganathan, S., 1966b, J. Less Common Metals **10**, 368.

Ranganathan, S., 1969, Phil. Mag. **19**, 415.

Ranganathan, S., K. M. Bowkett, J. J. Hren and B. Ralph, 1965, Phil. Mag. **12**, 841.

Ranganathan, S. and A. J. Melmed, 1966, Phil. Mag. **14**, 1309.

Ranganathan, S., R. C. Sanwald and J. J. Hren, 1966, Appl. Phys. Letters **9**, 393.

Rao, P. and G. Thomas, 1967, Acta Met. **15**, 1153.

Read, W. T., 1953, Dislocations in Crystals (McGraw-Hill, New York).

Read, W. T. and W. Shockley, 1950, Phys. Rev. **78**, 275.

Redmond, R. F., R. W. Klingensmith and J. N. Anno, 1962, J. Appl. Phys. **33**, 3383.

Rendulic, K. D., 1967, J. Less Common Metals **12**, 44.

Rendulic, K. D. and E. W. Müller, 1966, J. Appl. Phys. **37**, 2593.

Rendulic, K. D. and E. W. Müller, 1967, J. Appl. Phys. **38**, 2070.

Rendulic, K. D. and Z. Knor, 1967, Surface Science **7**, 205.

Rendulic, K. D. and E. W. Müller, 1967, J. Appl. Phys. **38**, 550.

Rose, D. J., 1956, J. Appl. Phys. **27**, 215.

Ruedl, E., P. Delavignette and S. Amelinckx, 1968, Phys. Stat. Sol. **28**, 305.

Ryan, H. F. and J. Suiter, 1964, Phil. Mag. **10**, 17.

Ryan, H. F. and J. Suiter, 1965, J. Sci. Instr. **42**, 645.

Ryan, H. F. and J. Suiter, 1966, Acta Met. **14**, 847.

Sanwald, R. C. and J. J. Hren, 1968, Surface Science **9**, 257.

Schultz, H., 1964, Acta Met. **12**, 761.

Schwartz, D. M., 1968, Ph. D. thesis, University of Cambridge.

Schwartz, D. M., A. T. Davenport and B. Ralph, 1968, Phil. Mag. **18**, 431.

Schwartz, D. M. and B. Ralph, 1970, Metal Science J. **3**, 216.

Seidman, D. N., R. M. Scanlon, D. L. Styris and J. W. Bohlen, 1969, J. Sci. Instr., Series 2, **2**, 473.

Sinha, M. K. and E. W. Müller, 1964, J. Appl. Phys. **35**, 1256.

Sleeswyk, A. W. and C. A. Verbraak, 1961, Acta Met. **9**, 19.

Sleeswyk, A. W., 1963, Phil. Mag. **8**, 1467.

Smith, D. A., 1968, Ph. D. thesis, Cambridge University.

Smith, D. A., 1969a, J. Sci. Instr. **2**, 106.

Smith, D. A., 1969b, Phys. Stat. Sol. **31**, 347.

Smith, D. A. and K. M. Bowkett, 1968a, 4th European Regional Conference on Electron Microscopy, Rome, 261.

Smith, D. A. and K. M. Bowkett, 1968b, Phil. Mag. **18**, 1219.

Smith, D. A., M. A. Fortes, A. Kelly and B. Ralph, 1968, Phil. Mag. **17**, 1065.

Smith, D. A., T. F. Page and B. Ralph, 1969, Phil. Mag. **19**, 231.

Smith, G. D. W., 1968, D. Phil. thesis, Oxford University.

Smith, P. J. and D. A. Smith, 1970, Phil. Mag. **21**, 907.

Smoluchowski, R., 1941, Phys. Rev. **60**, 661.

Southon, M. J., 1963, Ph. D. thesis, Cambridge University.

Southon, M. J., 1966, Field-Ion Microscopy, Eds. Hren and Ranganathan (Plenum Publishing Corporation, New York).

Southon, M. J. and D. G. Brandon, 1963, Phil. Mag. **8**, 579.

Southworth, H. N., 1968, Ph. D. thesis, Cambridge University; 1968, Scripta Met. **2**, 551.

Southworth, H. N. and B. Ralph, 1966, Phil. Mag. **14**, 383.

Southworth, H. N. and B. Ralph, 1968, Symposium on Field-ion Microscopy, Georgia Tech., Eds. Hochman, Müller and Ralph (Georgia Tech. Press).

Sproul, W. D. and M. H. Richman, 1968, Trans. AIME **242**, 1186.

Taylor, D. M., 1970, Ph. D. thesis, Cambridge University.

Tsong, T. T. and E. W. Müller, 1964, J. Chem. Phys. **41**, 3279.

Tsong, T. T. and E. W. Müller, 1966a, Appl. Phys. Letters **9**, 7.

Tsong, T. T. and E. W. Müller, 1966b, 13th Field Emission Symposium, Cornell University.

Tsong, T. T. and E. W. Müller, 1967a, J. Appl. Phys. **38**, 3531.

Tsong, T. T. and E. W. Müller, 1967b, J. Appl. Phys. **38**, 545.

Tsong, T. T. and E. W. Müller, 1968, Symposium on Field-ion Microscopy, Georgia Tech., Eds. Hochman, Müller and Ralph (Georgia Tech. Press).

Turner, P. J., 1967, Ph. D. thesis, Cambridge University.

Turner, P. J. and M. J. Southon, 1967, 14th Field Emission Symposium, Washington.

Turner, P. J. and M. J. Southon, 1969, 16th Field Emission Symposium, Pittsburgh.

Turner, P. J., P. Cartwright, M. J. Southon, A. van Oostrom and B. W. Manley, 1969, J. Sci. Instr., Series 2, **2**, 731.

Vitek, V. and F. Kroupa, 1969, Phil. Mag. **19**, 265.

Von Ardenne, M., 1956, Tab. der Elektronenphysik, Ionenphysik und Übermikroskopie, Vol. 1 (Springer, Berlin) p. 581.

Waclawski, B. J. and E. W. Müller, 1961, J. Appl. Phys. **32**, 1472.
Wasilewski, R. J., 1965, Acta Met. **13**, 40.
Wald, M. S., 1963, Ph. D. thesis, Cambridge University.
Whitmell, D. S., 1965, Ph. D. thesis, Cambridge University.
Whitmell, D. S., 1968, Surface Science **11**, 37.
Whitmell, D. S. and M. J. Southon, 1965, Adv. in Electronics and Electron Physics **22**, 903.


SHORT COURSES ON FIELD-ION MICROSCOPY
1966, Short Course on Field-ion Microscopy, University of Florida; Field-ion Microscopy, Eds. Hren and Ranganathan (Plenum Publishing Corporation, New York, 1968).
1968, Symposium on Field-ion Microscopy, Georgia Tech.; Eds. Hochman, Müller and Ralph (Georgia Tech. Press, to be published).

# AUTHOR INDEX

# SUBJECT INDEX

249